# LINEAR    ALGEBRA

Vaughn  D.    Brown

A  Comprehensive Study of  the  LINEAR FUNCTION concept that is Student Friendly but is still Rigorous. The Algebraic approach will be supplemented where possible by the use of Geometry.

AuthorHouse Publishers  2007

First published by AuthorHouse 12/03/07


ISBN: 978-I-4343-1341-6(sc)



Printed in the United States of America

Bloomington, Indiana


This book is printed on acid-free paper.

This textbook is dedicated to my grandson whose involvement
in  Linear Algebra and whose interest in my thoughts on this material
gave me a reason to write a complete text.

It is an extension of our previous enjoyable sessions of
working  together on mathematical problems and concepts.

Thank you, Brian.

# CONTENTS

Preface.                                                           i.

Chapter 1.     What is Linear Algebra ?                            1.

        1.0   Introduction                                        1
        1.1   The beginnings                                      1
        1.2   Two lemmas                                          2
        1.3   The addition identity theorem                       2
        1.4   Linearity                                           5
        1.5   Linear property                                     6
        1.6   Scalar product in derivatives                       6
        1.7   Scalar product theorem                              6
        1.8   Extended linear property                            7
        1.9   History                                             7
        1.10  Linear property                                     8
        1.11  Function definition                                 8
        1.12  Linear algebra the content                          10
        1.13  Linear algebra its essence                          11
        1.14  Linear algebra a study of linear functions          12
        1.15  Linear and line difference                          12


Chapter 2.     Arrows/Vectors.                                    13.

        2.0   Introduction                                        13
        2.1   Equivalent class                                    14
        2.2   Equal class                                         14
        2.3   Equivalence as equal                                15
        2.4   One-to-one correspondence                           15
        2.5   As equals                                           16
        2.6   Space                                               16
        2.7   Letters as arrows                                   16
        2.8   Arrow/vector                                        17
        2.9   Parallel vector                                     17
        2.10  Equal vector                                        17
        2.11  Free vector                                         17
        2.12  Position vector                                     18
        2.13  Directed line segment/ordered pair of numbers       18
        2.14  Magnitude of a vector                               19
        2.15  Equal vectors                                       19
        2.16  Projection                                          19
        2.17  Vector in terms of end-points                       20
        2.18  Definition of addition of vectors as a geometry      20
        2.19  Sum of more than 3 vectors                          21
        2.20  Theorem of addition of vectors as an algebra         22
        2.21  Could reverse 2.18 & 2.20                           23

# CONTENTS

2.22 Associative theorem for addition          23
2.23 Communitive theorem for addition          23
2.24 Zero vector                               24
2.25 Negative vector                           24
2.26 Scalar product                            24
2.27 Distributive theorem                      26
2.28 Another Distributive theorem              27
2.29 Associative theorem for products          28
2.30 Scalar product identity                   28
2.31 Product zero                              28
2.32 Standard 3-space unit vector              29
2.33 Standard n-space unit vector              29
2.34 Exercise                                  30
2.35 Angle of inclination                      31
2.36 Collinear vectors                         31
2.37 Linear test                               31
2.38 Exercise                                  32

Chapter 3.    Abstract Vector Space.                    33.

3.0 Introduction                               33
3.1 Definition of an abstract vector space     34
3.2 Axioms                                     35
3.3 Explantions of the axioms                  36
3.4 [-1] X                                     36
3.5 Spaces                                     36
3.6 General                                    37

Chapter 4.    Vector Space Theory.                      37.

4.0  Introduction                              38
4.1  Basic definitions                         38
4.2  Examples                                  40
4.3  Exercises                                 40
4.4  Basic definitions                         41
4.5  Linear transformation                     42
4.6  Definitions                               43
4.7  Normalizing                               44
4.8  Examples                                  44
4.9  A problem                                 45
4.10 Exercises                                 46
4.11 Theorems                                  46
4.12 Geometric problems in general             48
4.13 Geometric problems                        48

CONTENTS

Chapter 5.     Vector Analysis.                                53.

        5.0   Dot product                              53
        5.1   Theorems                                 55
        5.2   Projection problem                       57
        5.3   The Cauchy-Schwarz inequality            60
        5.4   Exercises                                61
        5.5   Verification proof                       61
        5.6   The triangle inequality                  61
        5.7   Notes about 5.3 & 5.6                    62
        5.8   Modified triangle inequality             63


Chapter 6.     Vector Analysis in 3-space.                     63.

        6.0   3-space                                  64
        6.1   Position vector                          64
        6.2   Scalar Product                           64
        6.3   Addition                                 65
        6.4   Magnitude                                65
        6.5   Dot product                              65
        6.6   Directional cosines                      66
        6.7   The inequalities                         66
        6.8   General                                  67
        6.9   Cross product                            67
        6.10  Theorems                                 68
        6.11  Geometry of the cross product            69
        6.12  Component form                           71
        6.13  Distributive theorem                     71
        6.14  Extended                                 74
        6.15  Examples                                 74
        6.16  Equations of a plane                     75
        6.17  Extended                                 76
        6.18  Dihedral angle                           77
        6.19  Example                                  78
        6.20  Equation of a line                       80
        6.21  Example                                  81
        6.22  Triple vector product                    83
        6.23  Construction                             84
        6.24  Triple scalar product                    84
        6.25  Skew lines                               87
        6.26  Distance between skew lines              91
        6.27  Field of complex numbers                91


Chapter 7.     Matrix Theory.                                  93.

        7.0   Introduction                             93
        7.1   General                                  93

# CONTENTS

7.2   Definition of a matrix            94
7.3   Row/column matrices              95
7.4   Square matrix                    96
7.5   Basic concepts                   96
7.6   Operations                       98
7.7   Demensions                       99
7.8   Subtraction                      99
7.9   Scalar product                   99
7.10  As a vector space               100
7.11  Standard basis                  100
7.12  Illustration                    101
7.13  Transpose                       102
7.14  Square matrix                   103
7.15  Sigma notation                  104
7.16  Product in general              106
7.17  Definition of a product         107
7.18  Example                         109
7.19  Geometry of a product           110
7.20  Properties of products          111
7.21  Distributive laws               113
7.22  Transpose theorem               114
7.23  Inverse matrix                  114
7.24  General                         115

Chapter 8.    Systems of Equations.           116.

8.0   Introduction                    116
8.1   System of equations             116
8.2   Augmented matrix                117
8.3   Elementary operations           117
8.4   Equivalency                     118
8.5   Elementary operations extended  119
8.6   Echelon form                    123
8.7   Rank                            126
8.8   Inverse matrix                  133
8.9   Definitions                     136
8.10  Theorems/example                137
8.11  Calculate inverses              138
8.12  Inverse of a product            139
8.13  Square systems                  140
8.14  A system to solve               141
8.15  Solutions                       141
8.16  Echelon method                  142
8.17  Existence of solutions          144
8.18  Inverse method of solution      145
8.19  Examples                        147
8.20  Homogeneous system II.          148
8.21  Non-homogeneous solutions       150
8.22  Exercises                       153

# CONTENTS

Chapter 9.   Determinants.      154.

| | | |
|---|---|---|
| 9.1 | Definitions | 154 |
| 9.2 | Evaluation - general | 155 |
| 9.3 | Minors/cofactors | 158 |
| 9.4 | Evaluation - nth order | 159 |
| 9.5 | Properties | 160 |
| 9.6 | Evaluation - specific | 161 |
| 9.7 | Example | 161 |
| 9.8 | Cramer's rule | 162 |
| 9.9 | Example | 162 |
| 9.10 | Other properties | 164 |
| 9.11 | Rank | 165 |
| 9.12 | Solving systems | 165 |
| 9.13 | Inverses | 167 |
| 9.14 | Adjoint | 168 |
| 9.15 | Diagonal matrix | 172 |
| 9.16 | Theorem | 175 |
| 9.17 | Example | 175 |
| 9.18 | Loose ends | 176 |

Chapter 10.   Linear Transformations.      178.

| | | |
|---|---|---|
| 10.1 | Introduction | 178 |
| 10.2 | Linear transformations | 179 |
| 10.3 | Example | 180 |
| 10.4 | Linear combinations | 180 |
| 10.5 | General | 180 |
| 10.6 | As a matrix product | 180 |
| 10.7 | Linear geometric operations | 184 |
| 10.8 | Basis | 187 |
| 10.9 | Algebra | 188 |
| 10.10 | Example | 189 |
| 10.11 | Examples | 190 |
| 10.12 | Multi-variable functions | 192 |

Chapter 11.   Eigenvalues and Eigenvectors.      194.

| | | |
|---|---|---|
| 11.0 | Introduction | 194 |
| 11.1 | Eigenvalues/eigenvectors | 194 |
| 11.2 | Illustrations | 195 |
| 11.3 | Note | 197 |
| 11.4 | Another approach | 197 |
| 11.5 | Example | 199 |
| 11.6 | Exercise | 202 |
| 11.7 | Polynomial equation | 202 |

# CONTENTS

11.8  Concepts                                    203
11.9  Inverse matrix                              203
11.10 Triangular Matrix                           203
11.11 Eigenvalues and Linear Transformations      204
11.12 Similar Matrices and Eigenvalues            205

The Appendix.                                      A1.

A1. The Derivative                                 A1.

A2. Irrational Numbers                             A2.

A preface is a written preliminary to a discourse; it gives the student some useful information about the discourse itself.

First of all the "editorial we" will be used throughout the whole text. It will be as if the discoveries are made by actions of both the author and the student working together.

If one looks at the table of contents of a few Linear Algebra texts in the library before attempting this course, one would get the idea that these texts are the study of vectors and matrices. This can become confusing to the student, because he may have had this kind of coverage in his calculus text. It is usually not made clear in such texts that those items were there to help the derivative and other concepts make Computations — calculations that could lead to the solution of many real world problems. This is not the use that we make of those items in this text. The different use of vectors and matrices will be made clear here from Chapter One on through out the whole text.

We will use concepts from the student's previous courses to help us develop the materials of Linear mathematics as if they were a part of this course. There are mathematical concepts that we will need and use also that rightly belong in other disciplines. Because the student may not have had such courses, such items will be explained in the text or in the Appendix. In all cases the definitions, theorems, operations, ... will be handled in manners all the way from informal to rigorous. The level will be chosen in terms of what is best to help the student understand the material and still be mathematically accurate. Some items that we consider to be important but are too much for the main body of a first course text will be found in the Appendix. At least a casual reading of this appendix is considered to be part of this course, and a more careful reading when there is a need for such.

The author has tried to write this textbook with the student in mind at all times. The main purpose of a textbook should not be to just develop mathematical concepts but rather to communicate such development to the student in such a way that the student understands the concepts. It is hoped that the student then will acquire enough skill in working with these concepts so that this skill can be used to solve the problems the student will face in the "real" world. Thus, it is very important that the explanations and the symbolic notation used in the text actually contribute to the student's understanding and not necessarily create an elegant piece of mathematical discourse that looks good to the professional mathematician.

At all levels there is considerable new memory work and new symbols to be mastered by the student. It is easy for a mathematician to get so lost in the creative process of forming new definitions , building new theorems, and developing powerful new symbolic notations that he forgets the writing is for beginners in this field. This author has put himself in the student's shoes and tried to avoid "snowing" the student with a rigorous axiomatic mathematics just because the final result looks less clumsy with its use of all inclusive symbols. As an example , there is no question that the Summation symbol saves a lot of horizontal space, but it is felt that beginning students find it easier to work with  "$a_1 + a_2 + a_3 +...+ a_n$" than with the Sigma Notation  $\Sigma\ a_i$ : i = 1 through n.

 Linear Algebra can become a formidable exercise in formal Logic. While some of the basics of Logic have to be used, these basics have been kept simple and extra effort has been made to make them understandable to the student who is taking this as a first course involving Logic.

Professional printers  have available a large selection of symbols that can be used to print the great number that are necessary in a text like this, but students can not easily script these special forms. This author has tried to avoid such special types as **BOLDFACE**, *Italics*, etc. He has taken great care in the use of the symbols of grouping; brackets [ ] , braces { } , and Parentheses ( ). They sometimes have other uses. This author feels that the student should be able to tell at a glance what is meant when symbolic notations are used and not have to spend time puzzling out the meaning.

A large amount of this mathematics lends itself to essay type proofs. This author believes that such proofs are hard to follow and not very convincing to most students; they will be avoided. As different from many writings in linear algebra, the material in Chapter One will lead the student naturally into the material of Chapter Two. The Chapter Two material will lead naturally into the material of Chapter Three. This pattern will continue through out the whole text. It will be clear to the student at every point of the text why he is studying some concept and why he is studying it now; every effort will be made to explain to the student what we are doing, how we are doing it, and why it is necessary to do this. Performing computations have been the main activity of the student in his calculus course, but this is not true here. Through Abstraction we will be looking at the Structure of the mathematics. Our basic methods will be a blending of the mathematics of the student's previous studies and the mathematics of pure logic. While we also will be working toward a final solution to problems, we also will be " looking under the hood" to study the beautiful machinery that caused the solutions to work.

While the basic number system used is that of The Field of Real Numbers, $x \in R$ , it is easy to expand the system to that of The Field of Complex Numbers.  Usually we will use a, b, ... to represent Constant, Real numbers and call them Scalars. The Variable, Real numbers will be the usual x, y, ... . The set of Integers will be symbolized by  I.

When the address of an element in an array is important we will use the double subscript; as in: $a_{34}$.  The first subscript refers to the number of the row of the location of this element and the second to the number of the column. Because subscripts are not a normal operation of many typing instruments, they are eliminated in cases where the address is not important. In this case we use any form that indicates that the elements are different numbers. We are more interested in the convenience to the student rather than in the "professional" appearance of the final result.

This text has been carefully designed to help make the transition from the relatively simpler mathematics of a lower division course to the abstract and more sophisticated mathematics of an upper division course.

The best way for a student to learn this material is for the student to try and rewrite each new concept in the student's own words. These notes can then be used to help the student understand new material and to prepare for tests.

## 1.0 Introduction

Mathematics is the ultimate way to solve problems. The solutions are concise and unambiguous --- the same for all who obtain them. Its methods are clear and relatively simple to apply. It is true that high levels of mathematical study require special abilities, but large numbers of students have been able to master a great amount of mathematics and even that which is beyond the elementary level. Is it any wonder then that mankind has turned to mathematics to find ways to solve the enormous problems that occur in the Real World? To do this we must find a mathematical model for a problem, solve the mathematics, and then translate the solution back into real world terms. Sometimes the model is only approximate, but we have good ways in mathematics to handle approximations. One of the best models for this purpose is called LINEAR. As different from most textbooks, we will explain the What and the Why of this nomenclature. One interesting way to find these answers is to use some material from the student's previous course.

## 1.1 The Beginnings.

1. In Calculus with the invention of the derivative of a function in the late 1600's came the proofs of certain theorems. One very important theorem for the use in the computations necessary in the Integral Calculus as well as the Differential Calculus is: the derivative of a sum is the sum of the derivatives. Or, $D(x_1 + x_2) = D(x_1) + D(x_2)$ where $x_1$ & $x_2$ are two variable forms from the set of real numbers .                    [Proof in the Appendix]

2. While this property of the derivative function was used over and over again through out the whole calculus, it is amazing for it is not true of some common functions like $(x + y)^2 \neq x^2 + y^2$ , $\sqrt{[x^2 + y^2]} \neq \sqrt{x^2} + \sqrt{y^2}$ , $\sin(x + y) \neq \sin x + \sin y$ , and so on.

3. The derivative function  [in the Calculus we took it to be a formula] is complicated; so let us look at the simplest type of function that we have studied and relate it to this property:

If $f(x) = m x$ where m is some real constant,

Is $f(x_1 + x_2) = f(x_1) + f(x_2)$ ?

4. now $f(x_1 + x_2) = m [x_1 + x_2] = m x_1 + m x_2 = f(x_1) + f(x_2)$ ; so the property is true of this function.

5. We have proved the Theorem: If a function is such that $f(x) = m x$  [m  a constant ],

$$\text{Then} \qquad f(x_1 + x_2) = f(x_1) + f(x_2)$$

6. In previous work by mathematicians it was seen that the converse of this theorem also is true and found to be very important. Its proof is the kind of proof that we will be using in this text  [complicated but not that hard to follow step by step]; the proof should be read through at least once.

1.2 Two Lemmas that we will need in the following proof.

A. Given:    $f(x)$ continuous for all $x \in R$ and f such that $f(x_1 + x_2) = f(x_1) + f(x_2)$

To Prove:  $f(x_1 + x_2 + \ldots + x_n) = f(x_1) + f(x_2) + \ldots + f(x_n)$.

Proof:

1. $f(x_1 + x_2) = f(x_1) + f(x_2)$

2. $f(x_1 + x_2 + x_3) = f(x_1 + [x_2 + x_3]) = f(x_1) + f[x_2 + x_3] = f(x_1) + f(x_2) + f(x_3)$

3. By Induction  $f(x_1 + x_2 + \ldots + x_n) = f(x_1) + f(x_2) + \ldots + f(x_n)$.

B. The result of dividing a Rational Number by a non-zero Integer is still a Rational Number:

1. now a/b where $a,b \in I$ & $b \neq 0$ is a rational number by definition,

2. and $bc \in I$ ; so a/bc where $a,b,c \in I$ & $b \neq 0$ , $c \neq 0$  is still Rational.

1.3 The Addition Identity Theorem

Given:   1. $f(x)$ Continuous for all $x \in R$

2. and f is such that $f(x_1 + x_2) = f(x_1) + f(x_2)$

To Prove: $f(x) = m x$  where m is some real constant and x a real variable.

Proof:  [ note: we will first prove this theorem for x a positive integer and then step by step extend x to all real numbers ]

I. Is there a Zero for this function ? We have seen in our work with functions in the real number system that the existence of a Zero in our system is very important; so: "does f(0) exist?".

>       1. f[0] = f[0 + 0] = f[0] + f[0]

>             = 2 f[0]                              note: the reasons are obvious

>       2. subtracting: 2 f[0] - f[0] = 0 or f[0] [2 -1] = 0

>       3. so for this function,    f[0] = 0

>       4. and now the answer is "yes" ; 0 is the Zero for this function f.

II.   use Mathematical Induction to prove the theorem for x a positive integer:

>       1.   f[x] is continuous for x e R so has a value at x = 1 ; label it m: so f(1) = m ,

but write it as : f[1] = m • 1 , where m is a constant

>             Note:

>                   a. f[2x] = f [x + x] = f[x] + f[x] = 2 f[x]

>                   b. let x = 1 then  f(2) = 2 • f(1) = 2 • m  or  f(2) = m • 2

>                   c. as far as a mathematical induction proof is concerned, steps a - b

are  not necessary; but, students feel much more comfortable starting this proof with x = 2 rather than with x = 1. We are not losing any generality nor rigor by showing this.

>             2. assume theorem is true for one case —the case where the independent variable is u :

>    or assume  f(u) = m • u    for a positive integer u.

>             3. examine the very next case ; or for the variable to be [u + 1];

>                f [u + 1] = f[u] + f [1]  by given

>                f [u + 1] = m u  + m

>                f [u + 1] = m  • [u + 1]    ;so true for the very next case from that case in step 2

>             4. thus f [x] = m x  is true for x = 1 from step 1;

>                and by step 3  true for x = 2 the next case;

>                and by step 3 is true for x = 3 the next case from that;

and so on for all positive integral values of x.

III. consider the theorem for negative integers, now let x>0:

1. $0 = f[0] = f[x - x] = f[x + (-x)] = f[x] + f[-x]$

2. $0 = f[x] + f[-x]$   implies $f[-x] = - f[x]$

3. $f[-x] = - m x f[-x] = m[-x]$   where $-x$ is a negative integer

4. so $f[x] = m x$ when x is negative.

5. theorem is now true for all integral x's;      or $f(x) = m x$ ,   $x \in I$

IV. Extend the Domain of x to the Positive Rational numbers. A rational number is the indicated division between two integers where the denominator is not zero. We will go no further into the theory of numbers than we have been before in previous mathematics.

1. Given the rational number y in the form of: $y = b/c$ where $b,c \in I$ & $c \neq 0$

2. f continuous for all $y \in R$; so it has a value at $x = 1$.  Label it m; so $f(1) = m$

3. or $m = f(1) = f(c \cdot [1/c]) = f(1/c + 1/c + \ldots$ to c terms $)$

4. from Lemma A.  $m = f(1/c) + f(1/c) + \ldots$ to c terms $)$
   $$= c \cdot f(1/c)$$

5. solving for $f(1/c)$:  $f(1/c) = m \cdot [1/c]$

6. from step 1: $1/c = y/b$ which is rational from lemma B; thus $f(y/b) = m \cdot [y/b]$

7. let $y/b = x$  any rational number;

8. thus $f[x] = m x$  for all Positive, Rational Numbers, x.

9. by the same argument as in III. we can extend this to the negative to get:
   $f[x] = m x$  for all Rational Numbers.

V. Extend the Domain of x to all Irrational Numbers:

1. This involves a deeper dip into the realm of Number Theory than we can cover complete; see Irrational Numbers in the Appendix:

a. Let an Irrational Number, b, be defined as a Nested Sequence of Rational Numbers,

b. let $\{r_n\}$ represent a Sequence of Rational Numbers; as:
   $\{r_n\} = r_1, r_2, r_3 \ldots, r_n$ where $r_k : k = 1,2,3, \ldots, n$  is Rational

c. While beyond the scope of this discussion [although this is made reasonable in the Appendix] , it can be shown that given any irrational number, b , then  where all Limits are taken as $n \to \infty$  we have    : $\lim r_n = b$

   2. so $f[b] = f( \lim r_n )$

               $= \lim f[r_n ]$        from Limit Theorems in the Appendix

   3.    $f[b] = \lim m \, r_n$

               $= m \lim r_n$

   4.    $f(b) = m \, b$

   5. thus:  $f[x] = m \, x$ for all possible Irrational Numbers.

VII.

   A.  To repeat; we have proved the very important theorem : The Addition Identity Theorem:

       Given : $f[x_1 + x_2 ] = f[x_1 ] + f[x_2 ]$ where f is Continuous for all x є R

       Then:   $f[x] = m \, x$  for all possible Real Numbers & m a constant.  Note: $m = f(1)$

   B. Students should be able to understand this proof, but the ability to duplicate the proof could be considered optional. The theorem itself is very necessary to our course.

1.4 Linearity.

   1. $f[x] = m \, x$  , m є R  &  a constant,  graphs into lines that pass through the Origin; so $f[x_1 + x_2 ] = f[x_1 ] + f[x_2 ]$ implies straight lines through point 0.

   2. These lines have geometric importance, but have additional great importance for our mathematics as we will see later.

   3. The equation form of all straight lines is   $a \, x + b \, y + c = 0$   which does not satisfy this property:

     a.   $a \, x + b \, y + c = 0$  implies    $y = f[x] = -\{a/b\} \, x - \{c/b\}$

     b.   where  u & v are different values of x

     c. $f[u + v] = -\{a/b\} [u + v] - \{c/b\}$

              $= -\{a/b\} \, u - \{a/b\} \, v - \{c/b\} = [-\{a/b\} \, u - \{c/b\}] - \{c/b\}$

              $= f[u] - \{c/b\}$    this is not  $f[u] + f[v]$  so not Linear;

however,  without changing the line we can use Analytic Geometry translation of axes

formulas x = x' & y = y' -c/b to change the form of the equation to y' = {a/b} x'.

These lines do pass through the Origin; so they do satisfy the property.

 4. As we will see later, we have a way to consider that all lines pass through the

origin without having to translate the axes; thus lines seem to  satisfy this linear

property.

5. It seems reasonable then to associate this property with straight lines and to use

the word LINEAR in its nomenclature. There are functions that do not have this

connection with the line f[x] = m x that also satisfy the identity

f[x + y] = f[x]+ f[y], but at the moment our prime consideration is the line

connection.

1.5 Definition of the LINEAR PROPERTY [ or LINEAR IDENTITY ] of a Function:

   A Function, f, is LINEAR if for any x in  the Domain of f ,

$$f[x_1 + x_2] = f[x_1] + f[x_2].$$

1.6.

   1. There was another theorem in calculus that went along with the one in 1.1.1

that we find interesting,

   2. D{f[s x]} = s D{f[x]}  where s is a constant, real number. See the Appendix for

the proof. In words we have the derivative of a function of a scalar product converts

into a scalar product of the derivative of that function.

   3. Products of this kind are special sums; as: if s is an integer,

      then   D{f(s x)} = D{f( x + x + ... to s terms)

      and    s D{f(x)} = D{f(x)} + D{f(x)} + ...to s terms.

   4. we see that there may be some kind of a connection between scalar products and The

Linear Identity.

1.7 Theorem.  If a function is Linear as in    f(x + y) = f(x) + f(y),

           then f(s x) = s f(x) where s is a constant and x is a Real number.

   Proof:

     1. f(x + y) = f(x) + f(y)          given

     2. f(s x ) = m [s x]               1.3. VII

     3.          = s [m x]              algebra

    4. f(s x)   = s f(x)                1.3. VII.

1.8 Linear definition extended:

    I. Because of the close association between the two forms, many writers choose both identities as the definition of a linear function. This avoids the proof of Theorem 1.3 but not the need for the theorem. One of the basic principles in the invention of a mathematics is to keep the number of hypotheses in a definition to a minimum; so we will keep 1.5 as the definition of linearity and use 1.7 as a theorem. However, for convenience we will many times list the two as an extended definition; as in:

    II. Extended Definition [used by some writers] for the LINEAR PROPERTY  or  LINEAR IDENTITY:

        Where f[x] is a Continuous function for x ∈ R  and s a  Real Constant,

        f is a LINEAR Function  iff :

        1. f[x + y] = f[x] + f[y]          called the ADDITION Property,  and

        2. f[s x]   = s f[x]               called the HOMOGENEITY Property

    III.

        A. It is easy to prove that the two properties can be written as one equation:

          The Function f has the LINEAR PROPERTY iff:

          f[a x + b y] = a f[x] + b f[y]  ,  x,y ∈ R  and  a,b ∈ R & are Constants.

          This is a further extended form.

        B. Exercise: Prove this extended form of the Linear Property using II.

1.9 A History.

    1. In the 1930's mathematicians became aware of a more abstract use of this property when applied to other functions for which it was true. By the 1960's this use became so powerful that a new course of mathematics was born. Study that started concentrating on the structure of mathematical things and abstracting the concepts involved lead to the structure/abstraction of other concepts as well --- some new and some old.

    2. Such a new course was labeled Linear Algebra usually with no explanation for the nomenclature. Colleges designed such courses, and textbooks were written to fit such courses. Even though the items covered, the definitions created, and the symbols invented have not been standardized to this day, there was one thing for which there was complete agreement. The Linear mathematics  was one of the most useful that a student could master

to apply to the solutions of the scientific and social problems of this modern world.

   3. The concepts that were used to tie 1680 - 1960 bits of mathematics together as a
discipline now seems to have actually created a powerful tool for the solving of many of
the problems of all of the sciences and the many other endeavors of mankind. We have found
that Linear Algebra helps us understand some of the Calculus processes and can blend the
divisions between algebra, geometry, and analysis into a single mathematical system. This
Algebra refers to a discipline where the emphasis is on How to perform operations on
objects that act like the real numbers,  Analysis refers to a study of Why those
operations obtain statements of truth, and  Geometry refers to the study of both the How's
and the Why's of problem solutions related to situations that can be related to geometric
figures. Algebra is the most commonly used name even by those who use a lot of Geometry to help
the student understand the material as we will do in this text.

   4. A part of the value of Linear Algebra will be found in the fact that later we will
have another way to look at the old techniques of solving systems of simultaneous, first
degree equations; these systems are a large part of our attempts to solve the problems
of the world.

1.10 The Property.

   1. It is clear this mathematics is a type of algebra/geometry; so some prefer the title
Linear Mathematics. Unfortunately most descriptions and the text in textbooks do not give
meaning to the qualifier, Linear. In some cases it would seem to have something to do with
lines, but a line connection to most of the mathematics is not given in those texts. In many
texts there is a mystery of why the word linear is used to partially name certain concepts.

   2. If the word linear is used in this text, it will be clear at all times why.

1.11 Definitions.

   Before we start the course, we have to take a step backward and define again certain
concepts from the past. Most of the concepts that we will use from previous courses have
been defined carefully enough that the intuitive sense that we bring to the discussions
will serve our purpose rigorously enough. The following , however, are important enough
that we have to make sure that we have something to fall back on when our intuitive sense
is in danger of leading us astray.

I.

1. Given two or more objects of some set that are connected is some fashion, then a RELATION is a complete description of how they are connected.

2. If A is related to B in some manner, we can symbolize this by writing  A ~ B where the symbol ~ is used to represent the Relation.

3. As: if △ ABC and △ DEF are such that angle A = angle D, angle B = angle E , angle C = angle F , then  △ ABC is similar to △ DEF where the Relation  [similar] can be written as △ ABC ~ △ DEF.

II.

1. As we use it here, a FUNCTON of X , as in Y = F(X), is a Rule which assigns to each X in some set S an unique value Y in S [ or some other set S' ]. For this course we do not have to be more rigorous in our use of the word "Rule". This clearly defines a Single-valued Function. All relations in this text are single-valued, and this is the only qualifier that we will make in this respect.

a. Set Theory is a whole, different mathematics whose totality of concepts are not necessary for this course. The intuitive feeling that a student will have when "X is an element of some set S" , [X є S] , is used will be sufficiently rigorous for this course.

b. The X is the Independent Variable of the function, and the total collection of X's that can be operated on by the Rule is the Domain of the function.

c. The Y is the Dependent Variable of the function, and the total collection of Y's  that are the consequences of the action of the Rule is the Range of the function.

d. Being overly careful and rigorous with certain qualifiers in our study tends to water-down the importance of the  main message of a definition, a theorem, or an explanation; so we will only be as rigorous as we have to to avoid contradictions, inconsistencies, ambiguities, and misunderstandings. It would be easy to go back and fill in any qualifiers that might seem necessary at some later date.

e. Equations will be thought of as Functions in some cases. As: a y + b = 0 implies y = -a/b and thus can be considered a Constant Function. Changing a x + b y + c = 0  into  y = -a/b (x) - c/b  allows us to think of the equation as a function of x.

III.  We will be using an "Operation" as another relation; so : An OPERATION on x,

as in y = O(x), is a Rule which operates on each x in some set S to get an unique value y

in S [ or S' ]. However, because of a possible confusion with the O symbol, we do not in

general use O for a relation symbol. We will use whatever is convenient like the T of this next

relation.

    IV.

        A TRANSFORMATION on x into y, as in T(x) = y, is a Rule which operates on each x in

some set S to transform it into an unique value y in S [or S']. Sometimes we call this a

Mapping.

    V. The definitions of function, operation, and transformation are basically the same so

the concepts can be used interchangeably. The choice sometimes is just a matter of

preference, or, which sounds the best. In y = $\sqrt{x}$  we are  more inclined to say that the

Square Root is an operation, and we are more inclined to think of y = sin x as a function.

1.12 Linear Algebra the Content.

    As seen in our above work, we will be doing what is clearly Algebraic;

however we also will use Geometry whenever its use adds to the understanding of our work.

Using Geometry increases the power of these concepts to study the real world and helps us

"see" the relationships.  The straight line is the most basic algebraic/geometric concept

with which we could work. Thus, it seems reasonable to use the words "LINEAR ALGEBRA" or

"LINEAR MATHEMATICS" label to any study of functions that satisfy THE LINEAR PROPERTY.

    The concept of linearity also is an important part of a serious study of other

concepts with which the student has had some contact in previous courses ; like: vectors,

matrices, and transformations [as in the rotation of coordinate axes]. We will see this as

we change our emphasis in the study of these concepts from the discovery of methods of

computation to the careful analysis of their structure in the abstract.

    A re-examination of previous work in Vectors and Matrices would show some

connection with idea of linearity as described above; so it is clear that we need to give

further attention to Vectors and Matrices but with the emphasis on structure. We will

relate our further study in this text to Linearity wherever it fits.

    Matrices are restricted to being finite dimensional [see later] spaces while

Vectors are not. Vectors methods are easier to use than matrix ones in many cases where both solution methods are available. A calculator is necessary in many matrix computations. These are some of the reasons this author has developed vectors before matrices in the next chapters.

These concepts and a Linear Transformation concept are what is taken to be the basic content coverage in what is usually a first course in LINEAR ALGEBRA. This text will cover more detail than is usual of each of these concepts plus some other higher level concepts.

To help us see the general plan of the content of this textbook we will look at one kind of summary of what we have done in our previous mathematics studies. We have invented and analyzed the following:

1. objects that can be used as elements of the domains of functions such as  the letters a,b,c,...,x,y,z representing Numbers such as 0,1,2,...,1/2,3/4,...  ;

2. their interactions with each other as with the operations Addition, Multiplication,...  ;

3. their fundamental properties as with Axioms, Postulates, Laws,…  ;

4. interrelations by the Function Concept as with $f(x) = m x$, $f(x) = \sin x$,...

5. all of this applied to the Solution of the Problems of Mathematics and the Problems of the Real World.

Prior to this course the Real Number System has been covered in much detail in terms of the above summary. What we will be doing in this course is to restrict our studies to functions that satisfy the linear identity, and we will extend our domains beyond the set of reals to sets of other objects that can be used in linear functions. We will study these new elements in terms of the five steps used above when we developed the real number system.


## 1.13 Linear Algebra its Essence.

While the content of a mathematics textbook seems to be filled with definitions, axioms, and theorems, the reading of the text many times seems to be like the watching of a magician pulling a rabbit then a flag then a bird out of an empty hat. The content seems to tell how it was done, but why this pass was a bird and not a flag is not clear. The certain motion on his part defined an end result that we call a bird, but we do not know Why it is this and not something else. In the previous courses of arithmetic, algebra, geometry, and even in the calculus the reason some definitions are stated in the manner that they are is not obvious. They seemed to be pulled out of a hat. In this text where the emphasis is on structure, the why should be a big part of the development of the material. This is what we have tried to do through out the text; this is the Essence of our Linear Algebra.

To give meaning to the actual mathematics chosen for the definitions we will abstract the basic definitions of the Real Number System to fit the various mathematical objects that we will study in this text. Here "Abstract" refers to a process whereby one recognizes the essential similarities between superficial different objects. We use the reals as a guide for our development because we are so familiar with that system , and that system has been so successful in working with numbers. We expect the same success in working with other objects.

1.14

With all of this work on a single concept it would seem that this course is an Algebra/Geometry whose sole reason for being is the study of Functions that satisfy the Linear Property; this involves the study of the Domains of Linear Functions, the Ranges of these Linear Functions, and the Rule itself that defines these Functions that are Linear.

This is what we will consider to be a course in Linear Algebra or Linear Mathematics. Note: the most common usage for the course is Linear Algebra.

1.15

Here is a final thought on what is Linear Mathematics in general. In most mathematics writings the word Linear is used too often in various, different ways; as in: a straight line and a first degree equation . This can be confusing to the student, because he does not know what the writer is referring to in this loose way when saying something is Linear. It is clear in this text that linear refers to The Linear Property. Later on we will see other concepts that use the modifier Linear in their names, but this is because the concepts are connected to The Linear Property.

When we are referring to the geometric concept of a straight line, we will say "line" rather than using the word linear. Note: in this text we do not study curved lines; so we can say "line" instead of straight line.

2.0 Introduction.

     As we have seen the important property of linearity is a concept of some functions and to study such functions we must have a set of objects [ the domain] for the "rule' of the function to work on. The set of real numbers has been covered in great detail in previous courses; so to expand the concept of linearity we must find other sets that might have some connection with the concept. Because of the close connection between linearity and straight lines we will first of all look at sets of objects that have a close connection with straight lines.

In physics we found that one of the most important concepts was that of a Force, and that a force, represented by a vector or arrow, was an action along a straight line. The basic connection of the all important Linear Identity to a straight line dictates a study of vectors in the abstract and in terms of this identity. Later we will find that the word vector also is used in a more general sense; so the emphasis in this chapter will on the arrow nomenclature even though both will be used because that is a common practice.

In this study of arrows we will emphasize the structure aspects of the concept. The little work that we did in the calculus with vectors involved finding sums and products of vectors by following geometric rules. There was no attempt to justify the rules. The student has had considerable experience in performing  addition in the Real Number System and maybe some work in justifying that rule; however, the rule for adding vectors is different from that for numbers. We call both of these operations Addition because they  both exhibit an additive effect. As, we start with some object, and then by some rule we operate on this first object with a second object of the same kind, and end up with third object of the same kind; thus the effect is additive.

     In later chapters we will be working with other objects beside arrows in this fashion; so it would become quite messy if we invented a different name for the operation every time we changed the objects. While some writers emphasize the difference by using different symbols like  +  &  ‡  , we will just use +  for addition no matter what the object is [ vector, matrix, ...].

We will call any operation that has a repeated additive effect with the same object a Product [or Scalar Multiplication ] and use the symbol  x, or · , or ( )( ),  or any other way that we used in algebra no matter what the objects are that are being multiplied. What we said above about addition , we can also say  about a product. Many of the other symbols that we use in the Real Number System or in our previous Algebra can be used here for concepts that are similar to those that were used in the previous systems. Here again it would be inconvenient to invent new symbols for all of those cases.

Because of our concentration on structure we have to use some concepts from the field of Logic, but we will only use those items that are necessary for the development of the idea of linearity. Some technical words we have to use with great care and study, but with others it is usually clear from the context of our work what is meant without further clarification. However, shortly there will be a time where more care in the description of our words will clarify what we are actually doing. First at this time we do need to discuss in some detail certain other concepts [2.1 – 2.7] that will act as a bridge between the algebra/geometry of the reals  and the algebra/geometry of arrows and the other objects of this course.

2.1 Definition of An EQUIVALENT CLASS.

   1. Given: a Relation ~ between objects that can be labeled W, X, Y, ..

      The  ~ is An EQUIVALENT RELATION and any element,W, is An EQUIVALENT CLASS,

         iff for any element W:

           a.  W ~ W ;                                    [called the Reflexive property]

           b. if W ~ X , then X ~ W ;                     [called the Symmetric property]

           c. if W ~ X   and  X ~ Y , then  W ~ Y.        [called the Transitive property]

      2. To better understand the case "a"   note that "great than"  can not be an equivalent relation because W > W is not true no matter what W represents.


 2.2 Because this sounds so much like "equals" we need a Definition of EQUALS.

    1. Given: a Relation = between objects that can be labeled W, X, Y, ..

       The = is an EQUAL RELATION  iff for any element W :

         a.  W = W ;                                     [called the Reflexive property]

         b. if W = X , then X = W ;                      [called the Symmetric property]

         c. if W = X   and  X = Y , then  W = Y.         [called the Transitive property]

d. if W = X, then W can be replaced with X in any mathematical expression without changing whether the expression is true or false.

2. It is clear that this is the EQUALS that we use in the Real Number system  and is the same as "is identical to".

3. The equivalent relation is a more general relation than that of the equal relation; similar triangles represent an equivalent relation but not an equal relation. These triangles fail the "d" part of the equals definition --- think of finding areas or perimeters.

4. All equal relations are equivalent relations, but not all equivalent relations are equal relations.


2.3 EQUIVALENT's as EQUAL's.

1. In order to build the mathematics that we are inventing here in this text, we need all of our "relations" to behave like they were Equal Relations. We need for the elements in our relations to be used like they were identical elements even though they are not identical. This is not unusual for while 2+3 is not "identical" to 5, the Arithmetic that we build and apply to the "real" world considers that they are in fact identical for they are identical in numerical value. This use fits one of the secondary definitions of the word "identical" as found in a dictionary.

2. We will not invent a special name or symbol for this special kind of equivalence. This is the same kind of thing that we discussed with addition.

3. So, if ~ is an EQUIVALENT RELATION, then W ~ X can be replaced with W = X in  all of the mathematical expressions in this text. We will use  the phrase "is identical to" as well as "is equal to" as the same thing through out this text.

4. We cover ourselves by saying that 2.3 is true by Definition.

5. The set of W, X, ... can be anything that can be tested in the definition--- like numbers, ordered pairs of numbers , vectors, ....

2.4 Definition of ONE-TO-ONE CORRESPONDENCE; symbolized by "1-to-1".

1. Given: B and C as two objects  and * is a relation between B and C such that:

a. for every C there is one and only one B, and

b. for every B there is one and only one C,

Then B * C is a ONE-TO-ONE CORRESPONDENCE between B and C.

2. Using the relation * as a 1-to-1 we will show that it is an EQUIVALENT RELATION that is also an EQUAL RELATION:

a. In the 1-to-1 definition change the C symbol to a B symbol for there is nothing in the definition that restricts how we label the second symbol; so for each B there is one and only one B. B is in a 1-to-1 relation with itself and thus B*B,

b. it is also clear from both sections of the definition of 1-to-1 that: if B * C, then C * B,

c. from 1-to-1: B transforms into one and only one, say, C; so C is now B. Now C, which is actually B, transforms into one and only one, say, D; so B is now D. And, there is no other way to get from B to D. This means that if B * C & C * D , then B * D,

d. because of the one-and-only-one aspect of both parts of 1-to-1, it is clear that we can satisfy the replacement hypothesis of the equals definition.

3. Thus 1-to-1 relations have satisfied all four conditions of an equals relation. We have now proved the following theorem.


2.5 Theorem: Given: the objects X, Y,... and a Relation ~. If any two of the objects are in a ONE TO ONE CORRESPONDENCE by means of ~ , then X is An EQUALS CLASS and the correspondence is an EQUALS RELATION.

Or, 1-to-1 in X & Y implies X = Y.


2.6 Quite often we will follow the common practice in Linear Algebra of calling a set of like objects a SPACE; so "a Set of Vectors" is "a Vector SPACE". While the big emphasis on the nomenclature "SPACE" starts in the next chapter, we will use it here also.

2.7

1. We will use capital letters like A,B,... to represent our arrows/vectors. These letters are sometime used in our dialogues as variables representing numbers. When there is a chance for confusion, we will use A ,B ,... with a "mark" over the letters as in $\bar{A}$ when A is an element of our space even if the A is not an arrow but definitely not a number. Using arrows as in this chapter, we will see that we are now blurring our algebra into a geometry. Seeing this graphic way of working with vector spaces will help a student understand   the concepts of Linear Mathematics/Algebra.

2.Every reference to a plane or to 2-space infers a Cartesian coordinate system in that space with (x,y) as the coordinates of the points.

3. To start the invention of the arrow concept we can use either a geometric or an algebraic development. It is easier to see reasons for the geometric definitions at the beginning level; so this how we will start each concept. Later it is easier to expand the definitions and then invent theorems by means of algebra; so that is how we will build the rest of the concept.

2.8. Definition of an ARROW [ or Vector ] in a plane, ( or in 2-space ).

I. An ARROW is Defined as a DIRECTED LINE SEGMENT in a plane,P ; so its Length [or Magnitude or Norm ] and its Sense of Direction completely determines the arrow.

II. Thus, an Arrow  is the straight line between, say, the point $B(b_1, b_2)$ and the point $C(c_1, c_2)$; this can be symbolized by BC. In some cases it is more helpful to think of  BC as a function that transforms the point B into the point C. The  Magnitude [some say: "the Euclidean Norm" ] of the arrow BC can be symbolized by $\|BC\|$. We can use |BC| if it is clear that we do not mean the "absolute value" of the number represented by the BC. Its direction is indicated by listing point B first  and C second. Because the initial point of the vector is not mentioned in the definition, the location of its initial point in the plane is immaterial in our definition. This is the most important aspect of the vector system that we will be using.

2.9 Definition of Parallel Vectors.

1. BC || DE if they satisfy all of the conditions of parallelism in Geometry;

2. this includes both vectors being on the same line.

2.10 Definition of Equal Vectors.

1. BC = DE iff BC || DE, |BC| = |DE|, & both have the same sense of direction along the line of direction.

2.11  Free Vectors.

Think of  all the  arrows/vectors that are equal to BC.  This set creates the EQUAL CLASS, BC.  All of the arrows are equivalent to each other and thus are also equal. So, Arrows can be moved about in space parallel to themselves and still be considered to be the same identical arrow.  A study based on this definition which sets up this equal class is the study of FREE VECTORS or FREE ARROWS, and this is the one we use. A Bound Vectors system is possible

but not as useful.

2.12. Definition of The POSITION VECTOR  in a Plane [or in 2-space]:      A $(a_1, a_2)$

   A.

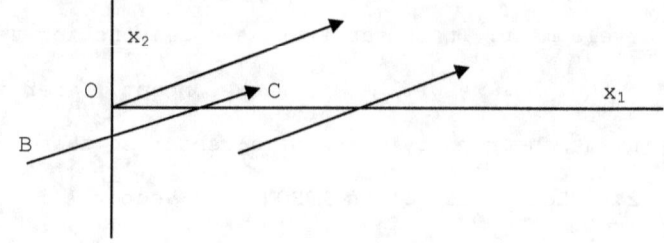

   1. Consider the equivalent class of BC

   2. The one member of the class whose

tail [or initial point] is on the Origin

is called the POSITION VECTOR of BC. Label the

Head [or terminal] point of the Arrow in this position, A $(a_1, a_2)$; so BC = OA. OA is the

Position  Vector of the whole class of vectors BC.

   3. Or [if it is more convenient] when looking at a certain arrow, we can consider that the

   Origin of a Coordinate System of this arrow is on it's initial point.

   B. An Arrow in this position is said to be in the STANDARD POSITION. Our fundamental

assumption in dealing with vectors is that unless stated otherwise  all vectors or arrows are

assumed to be in this position; so we can label vector OA as just vector A . If we are working

with the vector BC, we can use OA instead without applying the concept of free vectors or

indicating that we are dealing with a position vector.

   C. This position as Standard is partly dictated by the importance of our functions to be

   linear and thus related to Lines that do pass through the Origin. The following material

   gives further strength to our considerations of a Position Arrow as being the prime arrow

   of our class of arrows.

2.13. With the coordinates of point A being  $(a_1, a_2)$, we see that there is a one-to-on`e

correspondence between the vector A  and the ordered pair of numbers  $(a_1, a_2)$. For every

vector A there is one and only one ordered pair of numbers $(a_1, a_2)$ as the head point of

the vector and for every ordered pair of numbers $(a_1, a_2)$ there is one  and

only one vector A from the origin to that point. Because of the 1-to-1, A  is equivalent

and equal to $(a_1, a_2)$. This kind of equivalence converts a Geometric form into an

Algebraic form --- a directed line segment into an ordered pair of numbers. Thus,

BC = OA = A  = $(a_1, a_2)$ with the replacement possibility of A with $(a_1, a_2)$ and $(a_1, a_2)$

with A is of the greatest importance to our arrow mathematics. While this "is equal to"

is truly not the "is" of "identity", the above definition(s) allow us to use it as such.

Some writers use  ( , ) for Points and  < , > for Vectors or Arrows. We also will do this

[many writers use ( , ) for both].

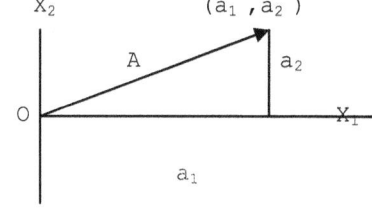

2.14. Using geometry we see that the

LENGTH [or MAGNITUDE or NORM] of vector A

can be evaluated by   $\|A\| = \sqrt{[a_1^2 + a_2^2]}$ ;

and if   $\|A\| = 1$ , then it is a  UNIT VECTOR [ or UNIT ARROW ].

2.15. More on equal vectors, A & B.

  I. where A = B:

   1.  Because of free vectors we can choose both vectors in the standard form--- that is

both initial points are on the origin,

   2. because they are parallel by geometry point B has to fall on the line OA,

   3. because of the equal lengths point B has to fall on point A,

   4. the two line segments coincide completely; so A = B.

  II. where   $A = <a_1 , a_2 >$  &  $B = <b_1. , b_2 >$:

.        a. then    $<a_1 , a_2 > = <b_1 , b_2 >$  iff  $a_1 = b_1$  and $a_2 = b_2$

         b. if A & B are position vectors this definition follows directly from I.

         c. if A & B are not position vectors, then they are equal if the first coordinates are

equal and if the second coordinates are equal  [note 2.17].

2.16 Definition of the PROJECTION of A on to B, symbolized by "Proj B A".

   1. Given: vectors A & B  as in figure:

Then Proj B A is the Vector formed by

dropping [PROJECTING] perpendiculars

from every point of A on to B;

so Proj B A  is a vector on B.

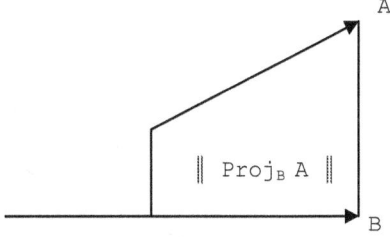

   2. Proj X A is the projection of A  on to the X-axis,

    and Proj Y A is the one on to the Y-axis.

   3. Given: P₁ (x₁ , y₁ ) & P₂ (x₂ , y₂ )

Then: $\| \text{Proj }_x P_1 P_2 \| = |x_2 - x_1|$ and $\| \text{Proj }_Y P_1 P_2 \| = |y_2 - y_1|$

2.17 Consider any vector AB where $A = (a_1, a_2)$ & $B = (b_1, b_2)$.

1. take OC = AB  and project on to the x-axis; so OD = EF  & DC = GB

2. by Geometry  $|EF| = b_1 - a_1 = |OD|$ which is the first  coordinate of OC

$|GB| = b_2 - a_2 = |DC|$ which is the second coordinate of OC

3. from step 1:

Theorem.  $AB = < b_1 - a_1, b_2 - a_2 >$ where $A = (a_1, a_2)$ & $B = (b_1, b_2)$.


2.18 ADDITION of TWO VECTORS  [or TWO ARROWS].

    I.

        1. In the real numbers the concept of addition was related to counting;
counting arrows would not have any meaning especially related to addition;

        2. thus we must look at this in some other manner.

        3. as we will see in the following, there is a geometric approach  that gives
the sense of addition directly.

    II. The Geometry.

        1. Given the vectors OA [or just A] and OB [or B] . Consider the sum : OA + OB.

        2. consider the figure:

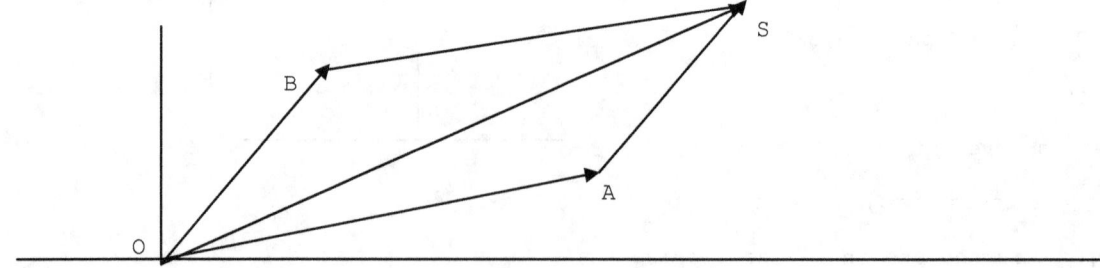

3. draw the vector AS = OB so that the tail of OB is on the head of OA,

    4. draw the vector OS

    5. By our Definition of the Addition of two vectors  OA + OB = OS. We will show that this definition is consistent with all of our vector concepts and with all of our impressions of the addition concept.

        a. consider moving a point from point O along the line OA to point A ,

        b. the vector OA clearly describes such a motion [correct length & correct direction long the line],

        c. now move the point from point A  along the line AS to point S ,

        d. the vector AS describes this motion,

        e. the Additive Effect of this motion is the same as if we had moved the point O along the vector AS to the point S,

        f. it is reasonable to feel that the vector OS is the result of the required addition, and we define it as such in the following:

    6. so  DEFINITION OF VECTOR/ARROW ADDITION of two vectors OA & OB is:

$$OA + OB \equiv OA + AS = OS \qquad \text{because } AS = OB$$

7. and the same fashion  $\overline{XY + YZ}$ = XZ ; RT + TS = RS ; and so on.

8. This is sometimes called the PARALLELOGRAM RULE, for the points O, A, S, B form a parallelogram with the DIAGONAL , OS, as the SUM of OA & OB.

9. The Sum exists and is unique  because we were able to do that Construction step-by-step. This  'able to do a Construction' is a good enough reason to satisfy all Existence/Uniqueness requirements in this text. We will not take the space to stress this in most cases, for we are not that deep into Logic Theory. We call this the CLOSURE LAW for the ADDITION of VECTORS.

2.19 The SUM of THREE or more Vectors.

    A. Given: Vectors SB, BC, CM, & SM

    1. It seems that we are looking at a figure where the tail of vector BC is on the head of vector SB, and the tail of vector CM is on the head of vector BC, and  the tail of vector SM is on the tail of SB:

2.  a.  Now  SB + BC = SC

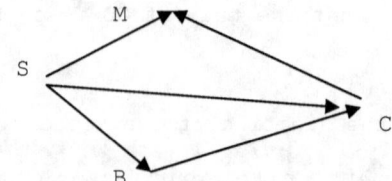

    b. and SC + CM = SM          B

   3. Thus it would appear that we are looking at the sum: SB + BC + CM = SM

 B. By extending this reasoning  we can find the sum of any number of vectors by

choosing a representative for each vector to fit the following labels:

    SM = SA + AB + BC + CD + ... + LM  . We will take this as our definition.

 C. Because the Head of the first vector is on the Tail of the second vector and so

on along the chain of vectors, some writers call this the Head-Tail method of Adding

Vectors.

 2.20 VECTOR ADDITION in terms of COORDINATES  --- The ALGEBRA.

   Note: we will be using the Equivalent/Equal concept to equate a vector and a point.

 Given : vector OA = < a,c >  &  vector OB = < b,d >

 Find by Proof:   the sum OA + OB = OS in terms of the coordinates.

 I. We will use geometry to prove the algebraic form of the Addition:

 II. In the figure we take AS = OB;  the rest below is obvious:

   1. OA + OB = OS    and OA = OE + EA , OB = OD + DB & OS = OF + FS

   2. |OB| = |AS|  and  angles DOB & CAS are equal; so rtΔ 's ODB & ACS are congruent

   3. so   |OD| = |AC| = |EF|   &  |DB| = |CS|  &  |EA| = |FC|

   4.        |OF| = |OE| + |EF| where |OE| = a  and |EF| = b; so |OF| = a + b

      &    |FS| = |FC| + |CS| where |FC| = c  and |CS| = d; so |FS| = c + d

   5. OS = <x,y>  where   x = |OF| = a + b   & y = |FS| = c + d

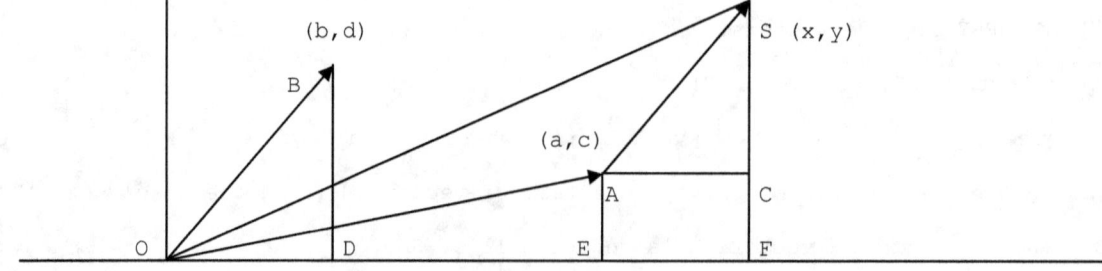

6. so OS = <a + b , c + d> ; now OA = <a,c>  & OB = < b,d>

7. so:   Theorem.   < a,c> + <b,d>  = <a + b, c + d>

2.21  We studied a Geometric situation that implied addition and invented a reasonable [2.18 (5)] geometric Definition for the Addition of two Arrows or Vectors. Using this definition we proved a Theorem that gave us an Algebraic form for the Addition of two Vectors. We  could start with an Algebraic Definition and then prove a Geometric Theorem, but there is no step 5 to make this definition reasonable.

2.22

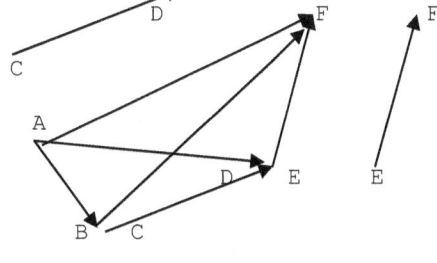

   Prove:  AB + ( CD + EF ) = ( AB + CD ) + EF

   1. AB + ( CD + EF ) = AB + CF = AF

      by vector addition,

   2. ( AB + CD ) + EF = AD + EF = AF

   3. so: AB + ( CD + EF ) = ( AB + CD ) + EF

   4. This is  The ASSOCIATIVE Theorem for ADDITION of VECTORS. Addition is done in

pairs, and it doesn't matter which association is made in the first pairing.

   5. algebraically: a. <a,b> +  [<c,d> + <e,f>] = <a,b> + [<c+e,d+f>] =<a+c+e,b+d+f>

                     b. [<a,b> + <c,d>] + <e,f>  = [<a+c>,b+d>]+<e,f>  =<a+c+e,b+d+f>

                     c. so;   <a,b> +  [<c,d> + <e,f>] = [<a,b> + <c,d>] + <e,f>:

2.23 Prove: AB + CD = CD + AB

   1. AB + CD = AD

   2. draw CE || AD; take |CE| = |AD|; draw DE

   3. ABED is a parallelogram; so

      DE || AB; |DE| = |AB|

   4. DE = AB; and from step 2.  CE = AD

   5. CD + DE = CE; so  CD + AB = AD

   6. from step 1. AB + CD  = CD + AB

   7. this is: The COMMUTATIVE Theorem for ADDITION of VECTORS. Changing the order of

adding two vectors does not change the value of the addition.

8. Algebra:<a,b> + <c,d>  =  <a + c, b + d> = <c + a, d + b>  = <c,d> + <a,b>

2.24 The ZERO VECTOR for Addition.

    I.   The ZERO VECTOR [ARROW] is defined as a vector with zero length and is written 0 or O or AA; so $|O| = 0$ or $|AA| = 0$. The direction of the zero arrow is not defined so has any direction that is convenient to choose at the time of its use. As a position vector, we have $O = <0, 0>$.

    II. Prove:    $AB + O = O + AB = AB$

      1. let $|AB| = a$ and we know that $|O| = 0$

      2. so $|AB| + |O| = a + 0 = a = |AB|$

      3.  O has any direction; so take $O \parallel AB$

      4. thus $AB + O = AB$ & $O + AB = AB$ from Theorem 2.23

    III.   We say that O is The ADDITIVE IDENTITY of Vectors. It is the Vector that does not change a given Vector under the addition operation.


2.25 The NEGATIVE of AB.

    I.

      1. The NEGATIVE of AB is BA or $-AB$; so $AB = -BA$.

      2. The negative of an arrow is the oppositely directed, but still parallel, form of the arrow; thus the negative of $<x, y>$ is $<-x, -y>$.

    II. Prove:    $AB + ( - AB ) = ( - AB ) + AB = O$

      1.  $BA = - AB$

      2. so $AB + ( - AB ) = AB + BA = AA = O$

      3. and $AB + ( - AB ) = ( - AB ) + AB$ from 2.23.

      4. thus $AB + ( - AB ) = ( - AB ) + AB = O$

    III. $( - AB)$ is the ADDITIVE INVERSE or The NEGATIVE of $(AB)$


2.26.  Definition of SCALAR MULTIPLICATION with ARROWS/VECTORS.

I.

    1. Think of $2 V$ as: $2 V = V + V$ ; so $2V$ is the adding of a vector to itself or a doubling of the Vector $V$ long the same line;

2. thus thinking of vectors being multiplied by scalars in general, it is

   reasonable to define such a scalar product as a TELESCOPING of the vector V by

   an amount equal to the value of the scalar [and along the same line]

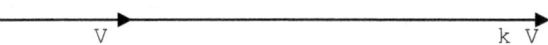

                       V                            k V

## II. Definition of a SCALAR PRODUCT.

1. [ k V ] is a vector where k V || V and |k V| = k |V|  ; if k>0 , the direction
is the same as V's  ,and  if k<0 , the direction is the opposite.

2. Because of the system of Free Vectors we can take k V as acting along the
same line as the line of V; we will do this in most cases.

## III. Prove: k AB Exists and is an Unique Arrow in a plane P.

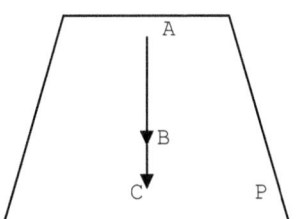

1. given AB in P and a scalar k in the Reals.

2. extend the line AB to point C

   so that |AC| = k |AB| = |k AB|

3. AC || AB ; so AC || k AB

4. AC = k AB ; so AC is an Unique Arrow in P,

5. thus: k AB Exists by construction and is an Unique Arrow in P.

6. This is called The CLOSURE Theorem for SCALAR MULTIPLICATION of VECTORS.

## IV. SCALAR PRODUCT in Coordinate form.

1. Theorem. Given: <x,y> a Vector in a 2-space,

       To Prove: k<x,y> = <kx,ky> where k is a scalar.

2. Proof:

   a. k<x,y> is a Scalar Product which is an arrow where k<x,y> || <x,y>

      so acts along the same line as <x,y> for both k>0 & k<0

   b. $|k <x,y>| = k|<x,y>| = k \sqrt{[x^2 + y^2]} = \sqrt{k^2} [x^2 + y^2]$

                 $= \sqrt{[k^2 x^2 + k^2 y^2]}$

                 $= \sqrt{\{[kx]^2 + [ky]^2\}}  =  |<kx,ky>|$  for both k>0 & k<0

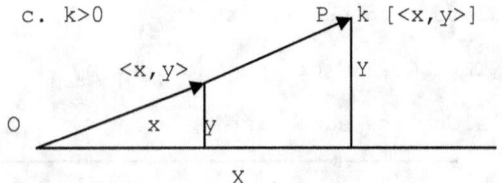

c. k>0                              P   k [<x,y>]              Y / |k <x,y>| =  y / |<x,y>|

        <x,y>              Y                                  Y = [ y • k |<x,y>| ] / |<x,y>|

    O        x     y                                         Y = k y

                   X

            X = k x  in the same manner

        d. so  k [<x,y>] = <k x, k y>          for both are the same vector OP

3. Exercise: as in step c. prove step c.    when k<0.

2.27  Prove:   k ( AB + CD ) = k AB + k CD

  I. As Vectors:

    1. AB + CD = AD

    2. extend the line AB to point E so that AE = k AB

    3. draw a line thru E || CD  ;  so  EF || CD ; thus  EF || k CD

    4. extend the line AD to cut the line in step 3 at point F;

     we now have the arrows AF & EF

    5. from geometry [similar triangles] & step 2:

       |EF| / |CD| = |AE| / |AB|    = k |AB| / |AB|= k

    6. |EF| = k |CD| by algebra; so EF = k CD

    7. from geometry [similar triangles] & step 2:

       AF| / |AD| = |AE| / |AB|     = k |AB| / |AB| = k

    8. so |AF| = k |AD| ; thus AF || k AD

    9. and AF = k AD

    10. now  AF = AE + EF

    11. using 2, 6, 8  into 9 : k AD = k AB + k CD

    12. from 1.                  k (AB + CD) = k AB + k CD

    13. This is called a DISTRIBUTIVE Theorem for SCALAR MULTIPLICATION of VECTORS.

  II. In terms  of coordinates:

    1.  r[<x,y> + <u,w>] = r[<x + u>, <y + w>] = <r[x + u], r[y + w]>

    2.   r[<x,y> + <u,w>] = <rx + ru, ry + rw> = <rx + ry> + <ru + rw>

    3.   r[<x,y> + <u,w>] = r<x,y> + r<u,w>

I. As Vectors:

1. take AB,BC,CD along line L as indicated; so AB||BC||CD||L,                A

2. k AB, m AB, & (k+m) AB are Arrows along L so are all parallel             B

3. let BC = k AB which means k of the AB arrows are being added one to

   the other [ Head-Tail ] along line L from pt.B to pt.C,                   C

4. let CD = m AB which means m of the AB arrows are being added one to

   the other [ Head-Tail ] along line L from pt.C to pt.D,

5. then BC + CD = [ kAB + mAB ] which means k of the AB arrows are

   being added one to the other [ Head-Tail ] from pt.B to pt.D and          D

   then from the tail,pt. C, m of the AB arrows are being added one to

   the other [ Head-Tail ] from pt. C to pt.D all along L in the same        L

   direction as  k AB

6. this actually means the [ k AB + m AB ] represents [k + m] of the AB arrows

   are being added one to the other along L from pt.B  to pt. D,

7. and this represents: [ k + m ] AB along L from pt.B to pt. D; so it

   has the correct direction along L of AB,

8. now |k AB| +|m AB| = |k| |AB| +|m| |AB| =|[ k + m ] AB|

   because all of these lengths are just Real Numbers

9. We have proved that: k AB + m AB = [ k + m ] AB by the definition

   of equal vectors,

10. thus    [ k + m ] AB = k AB + m AB

11. This is called a another DISTRIBUTIVE Theorem for SCALAR MULTIPLICATION,

    the one involving a Scalar Sum and a Vector.

II. In terms of Coordinates:

1. $[r + s]\langle x,y \rangle = \langle [r + s]x, [r + s]y \rangle$          th. 2.27

2.                $= \langle rx + sx, ry + sy \rangle$                        algebra

3.                $= \langle rx,ry \rangle + \langle sx,sy \rangle$           addition of vectors

4. $[r + s]\langle x,y \rangle = r\langle x,y \rangle + s\langle x,y \rangle$  th. 2.27

2.29  Prove:    k( m AB) = (k m) AB

    I. As Vectors:

      1. The Left side:

        a. m AB means m number of AB 's being added in the manner of 2.28

          along the line of AB,

        b. k (m AB) means k number of the  [ m AB ] 's being added as in 2.28

          along the line of AB,

        c. by the definition of a product this means (km) of AB's are being

          added as in 2.28 along the line of AB,

      2. The Right side: (km) AB means (km) number of AB's being added in the

        manner of 2.28 along the line of AB.

      3. thus: k( m AB) = (k m) AB

    4. This is called an ASSOCIATIVE Theorem for SCALAR MULTIPLICATION


   II. Exercise. Prove this theorem in the Coordinate form.

2.30. The SCALAR MULTIPLICATIVE IDENTITY.

   I. As Vectors:

      1.   1 • AB means we have just one AB in the "sum"

      2. thus: 1 AB = AB ; so we call    1 the SCALAR MULTIPLICATIVE IDENTITY.

   II. In Coordinate form:

      1. 1 • <x, y> = < 1 x, 1 y>

      2. 1 • <x, y> = <x, y>

2.31  The ZERO Property for a Scalar Product.

      1. 0 • AB means we have no or zero AB's in the "sum"

      2. thus clearly this is the ZERO arrow for we have no length

        and no defined direction; so

      3.   0 • AB = O = 0

2.32 The STANDARD UNIT VECTORS  i & j in 2-space.

1. Set up a coordinate system in the 2-space,

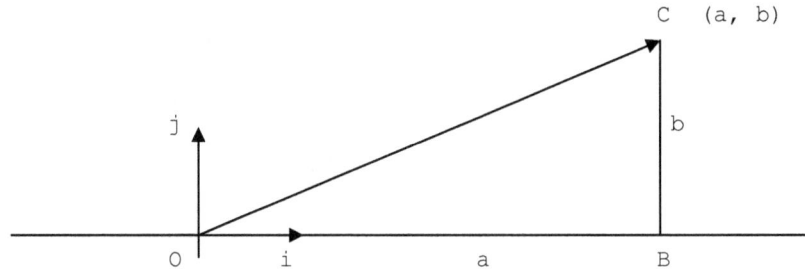

2. We will define two special vectors  i & j   in the 2-space:

a. vector i is the arrow from (0,0) to (1,0);   $|i|= 1$ [so a Unit vector]

and the vector i  is on the X axis. Or, vector i = $\langle 1,0 \rangle$ .

b. vector j is the arrow from (0,0) to (1,0);   $|j|= 1$ [so a Unit vector

and vector j  is on the Y  axis. Or, vector j = $\langle 0,1 \rangle$ .

3.   i & j have been used as the standard unit vectors for 2-space,

4. Look at the vector OC in the figure; from the previous definitions we see

that OC = OB + BC  , and a i = OB while b j = BC ;

5. so,    OC = a i + b j   .

6. It is clear that all position vectors, $V_2$ , can be expressed in the i/j form;

so, for point V ($v_1$ ,$v_2$ ) we have      $V_2$ = OV = $v_1$ i + $v_2$ j  = $\langle v_1 , v_2 \rangle$

7. In 3-space we take k = $\langle 0, 0, 1 \rangle$; then $V_3$ = $v_1$ i + $v_2$ j + $v_3$ k = $\langle v_1 ,v_2 , v_3 \rangle$

2.33  Definition of a Second form for STANDARD UNIT VECTORS .

1. We use both forms : i, j, k for 2-space & 3-space only , and $e_m$ for all spaces

[see later] ; note that for 4-space  and higher we lose a geometric interpretation.

2. $e_m$   : m = 1,2,3,...,n    where:

$e_1$ = $\langle 1,0,0,...,0 \rangle$ ; $e_2$ = $\langle 0,1,0,...,0 \rangle$ ;...; $e_n$ = $\langle 0,0,0,...,1 \rangle$

where the number of components depends on the size of the space.

3. So,  V = a $e_1$  + b $e_2$  where $e_1$ = $\langle 1,0 \rangle$, $e_2$ = $\langle 0,1 \rangle$.

...................................................................

V = a $e_1$  + b $e_2$  + ... + n $e_n$

2.34 Exercises.

  A. Given: points  B $(x_1, y_1)$  & C $(x_2, y_2)$

     Find: BC

     1. see figure for BC; draw OB & OC,

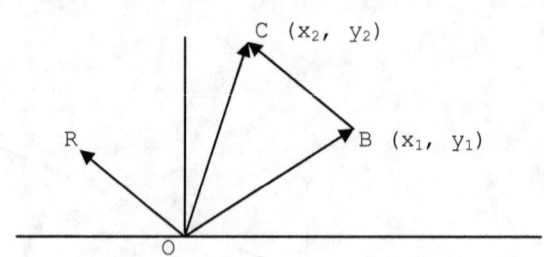

2. BC = BO + OC = OC + BO = OC − OB = $[x_2 i + y_2 j] - [x_1 i + y_1 j]$

    = $x_2 i + y_2 j - x_1 i - y_1 j = x_2 i - x_1 i + y_2 j - y_1 j$

  BC  = $[x_2 - x_1]i + [y_2 - y_1]j$   or   BC = $< x_2 - x_1, y_2 - y_1 >$

3. note:   $|BC| = \sqrt{[x_2 - x_1]^2 + [y_2 - y_1]^2}$


B. Find the Position Vector, R. of this BC.

 1. now  R = BC so R || BC  & |R| = |BC|

 2. perform projections as indicated in the next figure    '

 3.  angles DCB & FRO are = and angles CBD & ROF are =,

    because corresponding sides are ||. Thus Δ BDC is congruent to  Δ OFR. Then

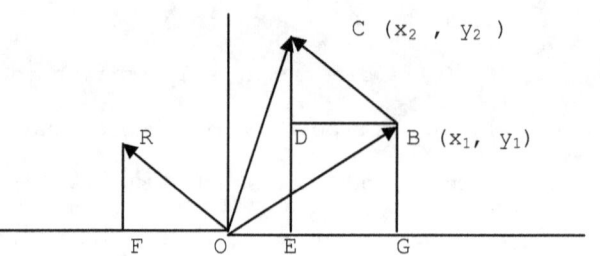

 4. $|OF| = |BD| = |GE| = x_2 - x_1$

and     $|FR| = |DC| = y_2 - y_1$

therefore  R = $[x_2 - x_1]i + [y_2 - y_1]j$

or      R = $< x_2 - x_1, y_2 - y_1 >$

 5. note the directions of the labels and the subtractions so that the signs are

correct. OF & FR are going in the right direction; $x_2 - x_1 < 0$    & $y_2 - y_1 > 0$.

2.35 . Find the component form of the Unit Vector U in terms of $\theta$ the angle of inclination [ $0 < \theta < \pi$ ]

    This is the angle between the vector and the X- xis.

    1. let $U = < u_1 , u_2 >$ where $|U| = \sqrt{[u_1^2 + u_2^2]} = 1$

    2. $\cos \theta = u_1 / 1$ or $u_1 = \cos \theta$

      $\sin \theta = u_2 /_1$ or $u_2 = \sin \theta$

    3. so $U = \cos \theta$ i + $\sin \theta$ j or $U = < \cos \theta , \sin \theta >$

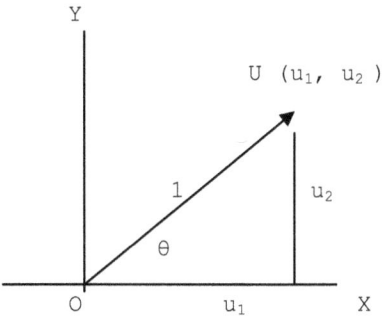

2.36 Collinear Vectors.

    1. Two vectors, V & W, that are on the same line [ collinear ] differ only in their magnitudes and in their directions along the line.

    2. so, $V = k W$ where $k \in R$       [ or $W = m V$ , $m \in R$ ]

    3. if V & W are Collinear, then $V = k W$ , $k \in R$,

      and, if $V = k W$ , $k \in R$ , then V & W are Collinear.

    4. because we use free vectors Collinear Vectors are Parallel Vectors.


2.37 Test for Linearity.

I.   1. Now that we have a new set of "elements" that can be used in the domain of a function we can see if there are functions of arrows that are linear.

    2. In all cases take the domain to be the set of all Position Arrows as in

    $A = < a_1 , a_2 >$ , $B = < b_1 , b_2 >$ ,    ...

    3. The fact that functions of ARROWS that are related to two or more independent variables instead of a single variable also can be LINEAR shows some of the additional power of the Linear concept.

II. Let F be a function that changes a vector A in a set S into k A where k is a positive constant. Is F Linear?

    1.  $F(B + C) = k [B + C] = k B + k C$

    2. $F(B) = k B$ and $F(C) = k C$ for all B,C $\in$ S

    3. thus $F(B + C) = F(B) + F(C)$ ; therefore F is Linear.

2.38 Exercises. Test the following functions for Linearity:

    A.   Let f be a function that reverses the direction of a vector. Hint: use the component form for the vectors.

    B.   Let f be the function that converts a positive vector into a vector that is the projection on the Y-axis.

    C.   Find the vector equation of the line L that passes through the origin with a fixed angle of inclination $\theta$ where $0 < \theta < \pi/2$.  Hint: Prove equation of L is:

        $f(r) = <r \cos \theta, r \sin \theta>$ where r is the magnitude of the vector A on L. Is f **Linear?**

3.0 Introduction.

We have just proved a set of theorems that defined a system of Vectors/Arrows. Even if the student has not had much study in modern algebra theory, he should recognize that these are the axioms or laws that define the Real Number System. Because these axioms might be basic as well to other systems of elements, it would be best at this time to generalize the concept before we do any further work in our vectors and the subsequent search for domains of possible linear functions.

We will build a structure that is an abstraction of those basic axioms that are used to define the Real Number System. All of the previous Definitions/Theorems are still true; some of the previous items are used to define a new Concept.

It is common to use the word Vector to name the basic elements that we will be using in additions and products. It would have been better to have chosen a more neutral name, but our language does not offer us much choice. So, for the moment we do not want the student to think of these vectors as Arrows or Directed Line Segments. The word vector is used here to describe an abstract element that satisfies the conditions that are set up in the list of axioms that we will be making; so the vector of this chapter is a somewhat vague quantity that is defined by not what it is , but rather by what it does. For the convenience of the student who is restricted to what can be hand-written, we will use symbols of this kind X, Y,... to represent the elements of our axioms what ever their meaning.

As is in common practice in this concept, we will call the Set of all objects that are of the same kind the Space, S, of those objects. We will use the word 'Abstract' to refer to a process whereby one recognizes the essential  similarities between  "superficial" different objects. When we have proved theorems about the objects of the abstraction, we are able to apply these theorems to the different objects in different spaces without further proof.

While in the following definition the basic elements can be different than numbers or arrows, we will still use + &  ·   for the general idea of addition and multiplication as we did in our study of arrows.

3.1 Definition of an ABSTRACT VECTOR SPACE over the Reals, R.

An Abstract Vector Space  consists of a nonempty set of Elements: X,Y,W,...

called Vectors, a set of real numbers  a,b,c,... called Scalars, and two operations:

Addition between Vectors, and Scalar Product between Scalars and Vectors.

Given the vectors X,Y,Z,  and the real numbers  a,b,c,        then

Addition [1-5]:

1.  X + Y   is an unique vector in the space ;            --- Closure Law

2.  X + ( Y + Z )  =  ( X + Y ) + Z ;            --- Associative Law

3.  X + Y  =  Y + X ;            --- Commutative Law

4. There is an unique vector  O  called the

Additive Identity [or Zero] Vector such

that for every Vector X we have that

X + O  = O + X  =  X ;            --- Zero Vector Law

5. For every vector X there is an unique

Vector  -X  called the Additive Inverse

[or Negative] of X such that

X + ( -X ) = ( -X ) + X  = O            --- Negative Vector Law

Scalar Multiplication [6 - 10]:

6.  a X  is  an unique vector in the space            --- Closure Law

7.  a ( X + Y )  =  a X  + a Y            --- a Distributive Law

8.  ( a + b ) X  =  a X  +  b X            --- a Distributive Law

9.  a ( b X )  =  ( a b ) X  = a b X            --- Associative Law

note: we can find the scalar product between X & b and then between that vector & a;

or we can find the product between X and the [arithmetic] product of a&b.

The results are the same; so the ( ) are not needed to show the association.

10.

I. There is an unique vector ,1, called One [Unit] such that for every vector X

1 X  =  (1)(X)  = X            --- Multiplicative Identity

II. There is an unique vector ,0, called ZERO such that for every vector X

    0 X  =  O                                                            --- Zero Law

      A. Proof:       1. X  =  1 X               Axiom 10

                  2. X  =  [ 1 + 0 ] X          arithmetic

                  3. X  =  1 X  +  0 X          Axiom 8

                  4. X  =  X  +  0 X            Axiom 10

                  5. 0  = 0 X                  Axiom 4

      B. We could consider that we are subtracting X from both sides; or Axiom 4 states that the only way a vector, 0 X, can be added to the vector X to get X is for the  0X to be zero.

      C. Some writers include this theorem as a second part of Axiom 10.


 3.2 Quite often we will use the common practice of shortening Abstract Vector Space to just Vector Space. Some writers consider that the identities listed in the definition as just part of the wording of the definition. They are the accepted Axioms of the Real Number system; so we will call them Axioms or Laws for our convenience and can use them in that fashion. All Definitions have an if-and-only-if [ iff ]feature; we can use this in the above. As: if a space satisfies all 10 of the laws, then it is an Abstract Vector Space; if it is a [Abstract] Vector Space , then all of the 10 laws are true.

 3.3  We said something about this before, but it will be well to again give a rough description of what these Laws mean. The Closure laws state that if you perform the operation, + or x ,  on  any two specific vectors of the space, you always get a vector of the space, and each result is always the same with the same two vectors [or the result is unique]. + and x is always an operation between two vectors. If you have to perform the operation on three vectors it has to be done in pairs. The Associative Laws state that it does not matter what order you do the pairing [ or what association you do first]. The Commutative Laws state that it does not matter which vector you start with first on the pairing. As we have seen in working with the Real numbers, we need a vector so that + [ or x ] gives us the same vector as the result of the operation.  We need a

vector so that + [ or x ] gives us a 0 as the result of the operation. Our present
algebra of numbers would not be possible without these concepts. So, our linear algebra
also needs them.

Some writers use  Z for the zero vector O , we will not at this time do this.


3.4 Theorem. ( -1 ) X  =  -X

I. Some feel that this theorem is so basic that it should be included in any
discussion of a Vector Space.

II. Proof:

|  |  |  |
|---|---|---|
| 1. | (-1)X + X  =  (-1)X + (1)X | Ax. 10 |
| 2. | =  (-1 + 1 ) X | Ax. 8 |
| 3. | =  0  X | arithmetic |
| 4. | =  O | Ax. 10 |
| 5. | (-1)X   =  -X | Ax. 5 |

3.5

I.  It is easy to prove that the Real Numbers are an Abstract Vector Space.

II. As an exercise do so.

III. The Theorems in Chapter Two that we proved for arrows are the 10 axioms of
the Definition of an Abstract Vector Space [AVS]; so the Space of Arrows is
An Abstract Vector Space. All of the theorems that we prove of Elements of an
Abstract Vector Space  will be true of arrows without any further proof --- assuming
that the basic elements make sense as arrows.

It is clear that for any function to satisfy the condition that
$f(ax + by) = a\ f(x) + b\ f(y)$ and  thus be linear its set of independent variables
would have to satisfy the theorems of a vector space. The theorems [ or Laws] of an
Abstract Vector Space give meaning to the additions and products that are necessary
in checking whether or not the linear identity is true for this function.

IV. Example: Is the function f( V ) = k V  Linear?

   1. f( V + W ) = k [V + W]

                = k V + k W   by step 7 of the AVS definition

   2. so: f( V + W ) = f( V ) + f( W )

   3. f( c V ) = k [ c V ]

          = c [ k V ]       by step 9 of the AVS definition

   4. f( c V ) = c [f( V )]

   5. so, this function of arrows is Linear; this is true only because the arrows

       are  elements of an Abstract Vector Space.

V. For a while we will be concentrating on the study of Abstract Vector Spaces where the word vector is used in its most general sense.  We will see then that the study of linearity is not restricted to numbers and arrows as the elements in the domain of our functions.


3.6     Generalizing the basic laws that define the Real Numbers we have generalized enough that this Abstract Vector Space could define other Spaces. We note that this is the case of the Space of Arrows as proven in Chapter Two. An Arrow might now be called a Vector because the item is now an all inclusive item indicating it is an element of a Vector Space. If we wish to emphasize the element is a directed line segment, we might call it an arrow and symbolize as such; otherwise we could use vector or even arrow/vector.

     It is clear that no function could be Linear unless its domain is an abstract vector space; the addition and scalar product parts of the definition of Linear need the truth of those parts of the vector space definition. So, while Linearity is the Foundation of Linear Algebra, an Abstract Vector Space is necessary for the existence of this fundamental basis of the course.

4.0 Introduction.

We will now expand our concept of vector spaces so that the word vector is used in the general sense. Sometimes in 2-space or 3-space we will use an arrow to help describe what we are doing. We will use Ă or <x, y, z> if we decide to emphasize the arrow interpretation of the vector. We have seen in the previous chapters that there is clearly a one-to-one relation between an arrow in the standard position and an ordered pair of numbers like (a,b) the coordinates of the terminal point A. When we study spaces other than arrows, we may find that the elements have more than two or three components.

The general, abstract vector V will now be defined as $V = (v_1, v_2, v_3)$ or in the general form of $V = (v_1, v_2, ..., v_n)$. Looking over previous definitions, theorems, and proofs, we see that we need only to add extra terms to expand into n-space. If we do expand beyond 3-space , we will lose our geometric or graphic way of describing our concepts. In some cases we might extend our geometric ideas by redefining these concepts to fit a n-space.

A certain set of general terms are used quite often in studying vector spaces at this level; so at this time we will define them and study some consequences.

4.1

A. Definition of a n-tuple.

1. The Ordered n-Component form $x = (a_1, a_2, ..., a_n)$ is called an n-tuple.

2. Thus, $A = (a,b)$ is a 2-tuple and $A = (a,b,c)$ is a 3-tuple.

3. Extending by induction the proofs of previous theorems  into n-space we see that the following are true:

a. $(a_1, a_2, ..., a_n) + (b_1, b_2, ..., b_n) = (a_1 + b_1, a_2 + b_2, ..., a_n + b_n)$

b. $k(a_1, a_2, ..., a_n) = (ka_1, ka_2, ..., ka_n)$

4. $O = 0 = (0, 0, ..., 0)$   is the  Zero  n-tuple.

B. Theorem. The Set , $V_n$ (R), of all n-tuples in the real number system is a

    Vector Space.

As a practice, run through this informally with 2-space. Some of the tests are so

simple they are almost trivial; so they are hard to write down. Symbols like

$V_n$ (R) indicating n components and over the Reals are not standard but are used by

some writers. This is true about a lot of the symbols used in Linear Algebra.

   C. Definition of a LINEAR COMBINATION.

    1.  If  X  =  $a_1$ X  + ... +  $a_n$ X  ,  then the sum, X, is called a

       LINEAR COMBINATION of the X  's with Coefficients  $a_k$  where k = 1,2,..

       This is also called the COMPONENT FORM of X.

   2. This can written by using the Sigma Symbol; as: X = $\sum a_k X_k$  where k = 1,2,..

The symbol is a space saver, but some students find the expanded form easier to

understand. We will use the expanded form in most cases.

   3. Simple example: if  A = 2 B + 3 C

             then A is a  Linear Combination of B & C

             and the 2 & 3 are the Coefficients.

   4. If we consider a Linear Combination as the result of an operation or

function, we would write it as T(x) = ( $a_1$  x  + $a_2$  x  +  ... $a_n$  x  )    thus

T(x + y) = ( $a_1$ [x  + y  ] + $a_2$  [x  + y  ] + ... + $a_n$  [x  + y ] )

       = ( $a_1$  x   + $a_1$  y   + $a_2$  x   + $a_2$  y   + ... $a_n$  x   + $a_n$  y  )

       = ($a_1$  x   + $a_2$  x  +...+ $a_n$  x  )+( $a_1$  y  + $a_2$  y + ... + $a_n$  y  )

       = T(x) + T(y)

T( c x ) = ( $a_1$  [c x  ] + $a_2$  [c x  ] + ... $a_n$  [c x  ] )

       = ( $a_1$  c x  + $a_2$  c x  + ... + $a_n$  c x  )

       = c ( $a_1$  x   + $a_2$  x   + ... + $a_n$  x  )

       = c T(x)

therefore this Combination is Linear by the Linear Identity of Chapter One;

this is the only reason for its name.

   5. The vector < 2, -1, 3 > = 2 i - j + 3 k  is a linear combination of the

standard unit vectors i, j, k.

D. Definition of LINEARLY INDEPENDENT.

1. The x's in the set  ( $x_1$ , $x_2$  , ... ,$x_n$  ) are LINEARLY INDEPENDENT if the

Linear Combination  $a_1 x_1$  + $a_2$  $x_2$   + ... + $a_n$  $x_n$  = 0

only if all the $a_k$ 's = 0: k = 1,2,...,n.

2. If we can find  $a_k$ 's  that make  $a_1 x_1$ + $a_2$  $x_2$ + ... + $a_n$  $x_n$  = 0 where least one

$a_k$  is Not Equal to  0, then the x 's in the set ( $x_1$ , $x_2$  , ... ,$x_n$  ) are

LINEARLY DEPENDENT

3. DEPENDENT means that we can solve for one of the x's in terms of the others.


4.2. Examples:

A. Consider a A + b B = 0 where A ≠ 0 & B ≠ 0

1. now      a A = - b B  which is true iff  a = 0  & b = 0

2. thus the A & B  are Linear Independent.

B. Given: X = <2,3> & Y = <4,4> is the set (X , Y ) Linear Independent?

1. b X + c Y = < 2b,3b > + <4c, 4c > = < 2b + 4c ,  3b + 4c > = 0 iff

2.  2b + 4c  = 0

3b + 4c  = 0  implies by subtracting   -b = 0 or b = 0

3. so in the 1$^{st}$ : 4c = 0  or  c = 0

4. therefore the X & Y  are Linear Independent.

C. Given: X = <2,3>, Y = <4,4>,  and W =<4,6> ; test this set.

1. b X + c Y + d W = < 2b+4c+4d, 3b+4c+6d > = 0  iff

2. 2b+4c+4d = 0

3b+4c+6d = 0  now subtact:  -b -2d = 0     or b = -2d

3. if we let c = 0 and let d = 1 ≠ 0 then b = -2 ≠ 0 ,

so both equations are satisfied; the X, Y, & W are Linear Dependent.

4.3 Exercises: Test whether the following sets of vectors are Linear Independent or

linear Dependent.

1. <1, -1, 2> , <2, -2, 4> , & <-3, 3, 1>

2. <1, -1> & <-2, 3>

4.4 There is a series of concepts that are tied together; so now we will discuss them as a package and go into the details later.

A. Definition of SPAN:

The set  S  of All linear  combinations, $X = a_1 X_1 + a_2 X_2 + \ldots + a_n X_n$ , as the a's vary through the real numbers is said to be the  SPAN of the $X_k$ 's  and is symbolized by  span( $X_1$ , $X_2$ , ..., $X_n$ ).  The  set S  is  Spanned or Generated by the  $X_k$ 's. The Span is a Vector Space, because each Linear Combination in the space is a Vector Space. Because of the length of the proof of this , we will wait until later to look at such proofs. We see that span(i,j,k) = a i + b j + c k , where a,b,c ∈  R, because every vector of 3-space can be generated by this form.

B. If $X_1, X_2, \ldots, X_n$ are Linear Dependent, then they can be spanned by some Proper Subset of the $X_k$ 's. For, at least one of the vectors X can be expressed as a linear combination of the others and thus be eliminated and still span the space.

C. Definition of DIMENSION.

If there is a finite set of vectors that Span a vector space  V, then V is said to be Finite Dimensional and n dimensional if there are  n independent vectors in the Span.  n is called the Dimension of the space V symbolized by: dim $V_n$  =  n; so dim $V_3$  = 3 and so on.

D. Definition of   BASIS.

1. If $X_1, X_2, \ldots, X_n$  span a space $V_n$  and the X 's are Linearly Independent, then the set ( $X_1$ , $X_2$ , ..., $X_n$ ) is a BASIS of $V_n$ = ( $X_1$ , $X_2$ , ..., $X_n$ ). So, a Basis of a Vector Space V is a set with the Minimum number of independent vectors in its Span.

2. A space can have more than one Basis.

3. In 4.2C  X = <2,3> & Y = <4,4> were proved to be Linear Independent. Can they be used as a Basis for 2 -space?

   a. can any vector V in 2-space be represented by a linear combination of the vectors X & Y?  and thus have V = a X + b Y    or:

      V = a<2,3> + b<4,4> = <2a,3a> + <4b,4b> = <2a+4b,3a+4b>,

   b. it is clear that 2-space is spanned by <u,w> where u,w ∈ R; so consider

      u = 2a + 4b       &   w = 3a + 4b

c. subtract: u − w = −a    or                                    a = −u + w

and        u = 2(w − u) + 4b = 2w − 2u + 4b    or        b = ¾ u − ½ w

d. so: for every u,w in 2-space there exists the coefficients  [ a,b] of a

linear combination of X & y; thus    [a X + b Y]     spans the 2-space,

e. the set <2,3> & <4,4> is a Basis for the 2-space.


## E. Definition of SUBSPACE.

1. If V is a Vector Space over R and U is also a Vector Space over R while being

a Proper Subset of V , then U is a Subspace of V.

2. The small amount of Set theory that we need here is that if every member of U

is also a member of V , the U is a Subset of A  --- that is U is included in V.

Proper refers to the fact that V contains at least one member that is not in U.

3. The set of all Rational numbers is Subspace of the set of all Real numbers.


## F. Definition of THE KRONECKER delta δ  ,   .

THE KRONECKER delta $\delta_{ij}$ , is a function such that:

$$\delta_{ij}     = 1   if   i = j   and$$

$$\delta_{ij}     = 0   if   i \neq j .$$

 at times it is convenient to use this symbol. Note: because of a different use we

should say that the i & j here are subscripts or indices [ not  arrows ].

1. we can write  $e_i$  = ( $\delta_{1j}$  , $\delta_{2j}$  , ..., $\delta_{nj}$ ) : i = 1,2,3,...,n   ;

So $e_1$  = ( $\delta_{11}$  , $\delta_{21}$ , ..., $\delta_{n1}$  ) ,..., $e_n$  = ($\delta_{1n}$ , $\delta_{2n}$  , ..., $\delta_{nn}$ )  ;

Thus   $e_1$  = (1,0,0,...,0), $e_2$  = (0,1,0,...,0), ..., $e_n$  = (0,0,0,...,1).

Where convenient we now can write Vector X as this n-tuple:

$$X = a_1  e_1  + a_2  e_2  +...+a_n  e_n   .$$

2. As an Exercise prove this X is a Basis for n-space.


4.5

We still have considerable work to do in developing the domains of functions that

could be linear, but there is a certain language we wish to use here and there that

involves saying a few more words [some a repeat] about the linear function concept before we start the real study of Linearity. At this level it is common practice to label linear functions as Transformations. Linear Transformations are Functions that Map Vectors in one Space into another vector in the same vector space [or in a similar vector space] while satisfying the Linear Identity.


A. Definition of a LINEAR TRANSFORMATION.

Let X and Y be two Vectors Spaces over the reals, R . Take T to be a Function that Operates on Vectors in X to TRANSFORM them into Vectors in Y such that [where $a \in R$ ; $x \in X$ ; $y \in Y$]:

$T(x_1 + x_2 ) = T(x_1 )+ T(x_2 )$   additive  property          [the definition]

$T(a \; x)$       $= a \; T(x)$              homogeneity  property        [a theorem]

Then  T is a LINEAR TRANSFORMATION  on X to Y; Symbolized by  $T: X \rightarrow Y$.
Note the change in the function symbol in this case; this is a common usage.

B. This transformation is also a LINEAR MAPPING and called a  Homomorphism [likeness in form]. We see the relationship between the LINEAR IDENTITY and a LINEAR TRANSFORMATION  and thus the name Linear.

C. The Linear Transformation Definition can be written in the single form:

$T(a_1 \; x_1 + a_2 \; x_2 ) = a_1 \; [T( x_1 )]+ a_2 \; [T( x_2 )]$   for  $T: X \rightarrow Y$.

Note: $T(x) = y$ for each x in the vector space X; so T is a Function of x.

D. As different from some of the discussion in Chapter One, we are interested now in Linearity more as an Operation or Transformation than as just some special function.

E. It is clear that this text is the study of Linear Transformations , but we save the complete study of this transformation as a Function for chapter 10.

4.6 Additional Definitions:

A. Definition of the KERNEL [or NULL SPACE] of a Linear Transformation.

Given: a $T: X \rightarrow Y$ and both X & Y are finite dimensional.  Then the set of all vectors, Xi :  i = 1,2,3,... such that  $T(Xi ) = 0$ is called the KERNEL [or NULL SPACE] of T and denoted by: ker T.  The set all vectors,Y, that come from $T(X) \rightarrow Y$, is called the RANGE of T and denoted by: rng T. The dim (ker T ) is called its Nullity and the dim (rng  T ) its Rank.

B. The Null Space or Kernel of the derivative, f', of a function f is the Set of

Constant Functions, f(x)= c,  because f'(c) ≡ D(c) = 0 .

   C. The Kernel of $D^{(n)}$ = $d^n$ f/$dx^n$   is the space of all Polynomials of Degree less than

(n-1). As an example: the Kernel of $D^{(3)}$     is the Space of all the

Polynomials: f(x) = a $x^2$   + b x  + c  and g(x) = d x + e and h(x) = k , because

f''' = 0 ; g''' = 0; h''' = 0.

   4.7 NORMALIZING A VECTOR.

   1. Given:    vector A = <$a_1$ , $a_2$ >  convert it into a UNIT VECTOR ,U.

   2. U  = A / |A| = A /  √[ $a_1^2$ + $a_2^2$ ]         where  |U| = 1  &  U || A


   4.8  Exercise: Given: any vector A = OA = a i + b j in 2-space

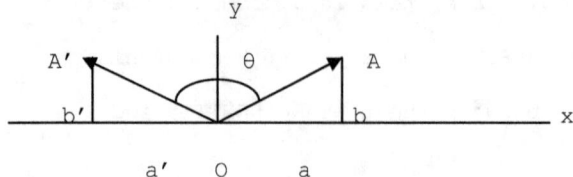

Rotate: A by an angle θ > 0 about the origin. Get: a vector of the form:

A' = a'i + b'j.  Find: A' in terms of a, b, &  θ ; then  Prove: the operation, rotation

R, of A is a Linear Transformation.  Or:  Prove: R:( a i + b j )  →  ( a' i + b' j )

is Linear.

   I. find (a', b') in terms of (a, b, &  θ ) where φ is the Angle of Inclination of A --

which is the angle between the X-axis and the vector.

   1. let |A| = c, then |A | = c and cos(Φ + θ ) = a'/c and sin(Φ + θ ) = b'/c

   2. a' = c cos(Φ + θ) = c cos Φ  cos θ - c sin Φ  sin θ

      b' = c sin(Φ + θ) = c sin Φ  cos θ + c cos Φ  sin θ

   3. cos Φ  = a/c  or a = c cos Φ

      sin Φ  = b/c  or b = c sin Φ

   4. in 2: b' = b cos θ + a sin θ and a' = a cos θ  - b sin θ

   5. or  b' = a sin θ + b cos θ   and     a' = a cos θ - b sin θ

6. We have proved that Rotating any vector V = x i + y j in 2-space about the origin by a positive angle $\theta$ gives the transformation:

T: (x i + y j $\rightarrow$ (x cos $\theta$ - y sin $\theta$, x sin $\theta$ + y cos $\theta$ )

II.  Now prove R is Linear using the two vectors A = ai + bj  and D = di + ej

1.

a.  R:A $\rightarrow$ A' is R:(ai + bj) = (a cos $\theta$  -  b sin $\theta$)i + (a sin $\theta$ + b cos $\theta$)j

b.  R:D $\rightarrow$ D' is R:(di + ej) = (d cos $\theta$  -  e sin $\theta$)i + (d sin $\theta$ + e cos $\theta$)j

c. sum: R:A + R:D=[(a+d)cos $\theta$  - (b+e)sin $\theta$]i + [(a+d)sin $\theta$ + (b+e)cos $\theta$]j

d. R:(A+D) = R:[(ai+bj) + (di+ej)] = R:[(a+d)i + (b+e)j]

so to find R:(A+D)in terms of : in II.1.a replace all a's with (a+d) and the b's with (b+e) to get:

e. R:(A+D)=[(a+d)cos $\theta$  - (b+e)sin $\theta$ ]i + [(a+d)sin $\theta$ + (b+e)cos $\theta$ ]j

f. so from c & e:   R:(A+D) = R:A + R:D   ; thus R satisfies the Additive Property.

2. Note:

a. k (R:A ) $\equiv$ k R:(ai+bj)= k[ (a cos $\theta$ - b sin $\theta$)i + (a sin $\theta$ + b cos $\theta$)j ]

= (ka  cos $\theta$  - kb sin $\theta$)i + (ka sin $\theta$ + kb cos $\theta$)j

b. R:k A $\equiv$ R:k(ai + bj) = R:(ka i + kb j)

= (ka cos $\theta$  -  kb sin $\theta$)i + (ka sin $\theta$ + kb cos $\theta$)j

c. so:  R:(k A) = k( R: A)     ; therefore the Homogeneity Property is satisfied.

3. therefore:   R:(A) $\rightarrow$ A' is a Linear Tranformation.

4.9  A problem: Given: pt.A (2,3) and pt.B (4,1)

Find: P the Standard form of AB & a Unit vector,U, along AB.

1.  Draw AO and OB and show  AB  in  the  coordinate  system,

2.  from the definition and geometry it is clear that:

a. AB  = AO + OB  = OB + AO = OB - OA

b. AB = 4i +j-[2i +3j] = 4i+j -2i-3j  = [4-2]i + [1-3]j

c. AB  = 2i -2j; note <2,-2> is the Position Vector P of AB.

d. |AB| = $\sqrt{(2)^2 + (-2)^2}$  = $\sqrt{4 + 4}$ = $\sqrt{8}$ = 2 $\sqrt{2}$

e. U = 2/ {2 $\sqrt{2}$}  i  -  2/ {2 $\sqrt{2}$}  j

f. or  U =  {1/$\sqrt{2}$} i  - {1/$\sqrt{2}$}  j

4.10 Exercises.

A. Express the UNIT VECTOR, $U_\theta$ , in terms of $\theta$ the Angle of Inclination which is the angle between the X-axis and the vector.

    1. $U_\theta = <u_1 , u_2>$ where $|U_\theta| = 1$

    2. $\cos \theta = u_1 /1$ so $u_1 = \cos \theta$ & $\sin \theta = u_2 /1$ so $u_2 = \sin \theta$

    3. thus $U_\theta = <\cos \theta , \sin \theta> = \cos \theta \; i + \sin \theta \; j$ where $|U_\theta| = 1$.

B. Express any Vector A in terms of $U_\theta$ .

    1. For any Scalar k : $A = k U_\theta = <k \cos \theta , k \sin \theta>$

    2. Or          $A = k \cos \theta \; i + k \sin \theta \; j$


4.11 Basis Theorem.

A. Given: the n-dimensional Vector Space, **V** ,

        with the Basis $B = ( B_1, B_2, ..., B_n )$

   Then: every V in **V** can be written in one and only way as a

        Linear Combination of the Basis [or the Combination is UNIQUE].

B.   Proof: 1. Given $V = a_1 B_1 + a_2 B_2 + ... + a_n B_n$ or $\sum a_k B_k$ : k = 1, 2, ..., n.

        2. assume the V can be spanned by a second Linear Combination

          of the Basis; as:

              $V = c_1 B_1 + c_2 B_2 + ... + c_n B_n$ or $\sum c_k B_k$ : k = 1, 2, ..., n.

       3. subtract: $0 = [a_1 - c_1] B_1 + [a_2 - c_2] B_2 + ... + [a_n - c_n] B_n$

       4. so    $a_1 - c_1 = 0$             $a_1 = c_1$

           $a_2 - c_2 = 0$             $a_2 = c_2$

           ...............................

           ...............................

           $a_n - c_n = 0$             $a_n = c_n$

       5. The Linear Combinations are the Same; so UNIQUE

C.   Exercise. Given: Set of Vectors, V , in a Vector Space $\mathbf{S}_3$ $\epsilon$ (R) where

        $V_1 = <1, 0, -1>$, $V_2 = <0 , 0 , 1>$ , $V_3 = <0 , 1 , -1>$.

        And, in the same Space $X = <2 , 3 , 0>$ & $Y = <1 , -1 , 1>$.

  Then: is the set of V 's a Basis for V? Now express X & Y in terms of this Basis.

1.  $c_1 V_1 + c_2 V_2 + c_3 V_3 = \langle c_1, 0, -c_1 \rangle + \langle 0, 0, c_2 \rangle + \langle 0, c_3, -c_3 \rangle$

$= \langle c_1, c_3, -c_1 + c_2 - c_3 \rangle$ which $= 0$  or  $\langle 0,0,0 \rangle$

iff $c_1 = 0$ & $c_3 = 0$   &   $-c_1 + c_2 - c_3 = 0$   where   $c_2 = 0$

2. so the set V is Linear Independent.

3. dim V = 3 for a 3-space Vector Space.

4. now does the set $V = c_1 V_1 + c_2 V_2 + c_3 V_3$  Span the Vector Space , $\mathbf{S_3}$ ? :

 a. take $A = \langle a,b,d \rangle$ as any vector in $\mathbf{S_3}$ where $a,b,d \in$  R,

 b. can A be created from $c_1 V_1 + c_2 V_2 + c_3 V_3$  ?

 c. Yes for in #1 let $c_1 = a$, $c_3 = b$, &

    then  $-c_1 + c_2 - c_3 = -a + c_2 - b = d$ or $c_2 = d + a + b$

 d. $c_1 V_1 + c_2 V_2 + c_3 V_3 = a \langle 1,0,-1 \rangle + [a + b + d] \langle 0,0,1 \rangle + b \langle 0,1,-1 \rangle$

$= \langle a,b,d \rangle$    ; so V scans $\mathbf{S_3}$ ;

5. therefore   $V = c_1 V_1 + c_2 V_2 + c_3 V_3$ is a BASIS.

6. from $X = \langle 2,3,0 \rangle$   let $c_1 = 2$, $c_3 = 3$, $c_2 = 0 + 2 + 3 = 5$

  And  $Y = \langle 1,-1,1 \rangle$  let $c_1 = 1$, $c_3 = -1$,  $c_2 = 1 + 1 - 1 = 1$

7. so   $X = 2V_1 + 5V_2 + 3V_3$    &  $Y = V_1 + V_2 - V_3$  .

4.12 Geometry problems.

  1. Vectors can be used to solve some interesting geometric 2-space problems. Before we take the next step into higher spaces we will look at a few. One method of finding solutions to complicated  Vector problems is described next.

  2. There is a technique to discovering geometric proofs. To prove these start by drawing the given geometry, and then draw in the conclusion. Then the geometric sense that was developed in a geometry course should indicate the method of proof --- as you will see by looking at the following proofs. If the reason for a statement is a simple, routine geometric definition or theorem, it probably will not be given. We will be using the words Vector and Arrow interchangeably.

  3. In all of the following theorems the Vectors are assumed to exist and be completely defined in some plane, P, or 2-space. The lower case letters like k,m,... are Scalars and thus are elements of the Real Number System, R.

4. Many solutions are in the form of a CARTESION EQUATION [ y = f(x) ], or

PARAMETRIC EQUATIONS [ x = f(t) & y = g(t) where t is the Parameter ], or the

COORDINATES of a POINT [ (x,y) ].

    5. Method: a. set up a convenient Coordinate System,

                b. find a general point P(x,y) on the curve or line or ...,

                c. find the Position Vector OP = x i + y j,

                d. by means of the Addition of Vectors find another path that gives OP

as the Resultant - one that involves distances and direction that you can work out,

                e. simplify step d,

                f. equate steps c & e to get the final solution.

    6. Can change Parametric Equations to the Cartesian form  by solving both

equations for the parameter and equating the two expressions.

4.13. Exercises.

I. Find the coordinates of the Centroid of the  triangle formed by points

    A(1,-2),B(3,4),& D(5,1).

    1. Draw the triangle ABD with loose accuracy; label the midpoints of AD as M'(x',y')

and of AB as M'' ( x'' , y'' ) making  no attempt to place points M' & M'' in their

correct positions; draw Medians BM' & DM''; label the Centroid C (x, y ).

    2. AD = (5-1)i + (1+2)j = 4i + 3j

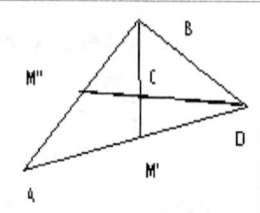

    3. M' is midpoint of AD; so AM' = ½ AD

$$= 2i + 3/2 \ j$$

    4. also AM' = (x' -1)i + (y' + 2)j      so    x'  -1 = 2        or      x'  = 3

                                       y'   +2 = 3/2                y'  = -1/2

    5.  BM' =(3 - 3) i + (-1/2 - 4) j   =  9/2 j

    6. from geometry: |BC| = 2/3 |BM'|    so BC = - 3 j and BC = (x - 3) i + (y - 4) j

7. thus    x - 3 = 0            and    x = 3

           y - 4 = -3                  y = 1

8.  Coordinates of the Centroid:   C(3,1)


II.  Given a circle with center (0,0) and radius r. A string is wound around the

circle, and then unwound while being held taunt in the plane of the circle. Let the

initial position of the tracing point P be A(r,0), and point T be the point of

tangency of the string. Find the vector OP that describes the motion of P in terms

of the angle AOT , $\Phi$ . From this we can get the  Parametric Equations of the curve

described by P. This is an  Involute of the circle.


    1. OP = x i + y j      also    OP = OT + TP    thus   OT + TP = x i + y j

    2. arc AT = r  $\Phi$    [$\Phi$ = angle AOT]   also arc AT = |TP|   the length of the string

unwound    thus   |TP| = r  $\Phi$

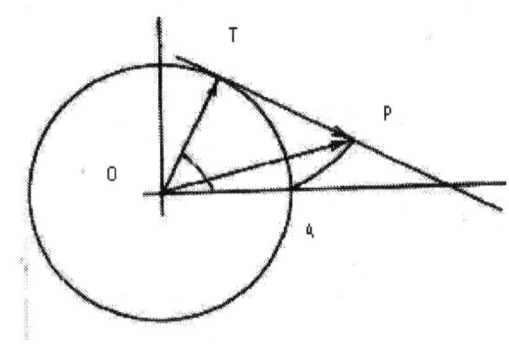

    3. OT = r U$_\Phi$   = r cos $\Phi$  i + r sin $\Phi$  j

    4. $\theta = \pi + \Phi + \pi/2 = 3\pi/2 - \Phi$

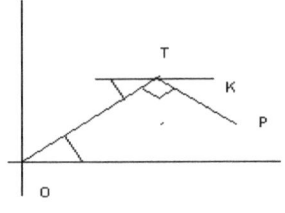

5. $\sin \theta = \sin(3\pi/2 - \Phi) = -\cos \Phi$          [θ = angle KTP]

   $\cos \theta = \cos(3\pi/2 - \Phi) = \sin \Phi$

6. $TP = r \Phi \quad U_\theta = r \Phi [\cos \theta \, i + \sin \theta \, j]$

   $= r \Phi \sin \Phi \, i - r \Phi \cos \Phi \, j$

7. steps 3 & 6 into 1:

   $x \, i + y \, j = r \cos \Phi \, i + r \sin \Phi \, j + r \Phi \sin \Phi \, i - r \Phi \cos \Phi \, j$

8. so:   $x = r \Phi \cos \Phi + r \Phi \sin \Phi$          The Parametric Equations

   $y = r \Phi \sin \Phi - r \Phi \cos \Phi$          of the Involute of the Circle.

III. Find the parametric equations of a point P on the circumference of a
   wheel [ circle ] as it rolls along a horizontal straight line without slipping.
   Work out in terms of the angle of rotation of the Y-Axis spoke [the line from
   point (0,0) to (0,r) the center of the first circle ] as Point P goes from (0,0)
   to (x,y). This curve is a Cycloid. Let C' be the center of the second circle.

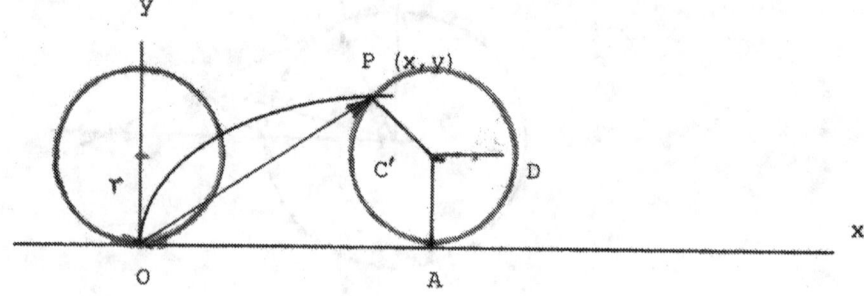

1. Draw the vector OP ; so OP = x i + y j  but also    OP = OA + AC' + C'P

2. if we roll the right hand circle back, the pt.P will fall on pt.(0,0). The arc
   PA will lay down on the line OA; so  |OA| = arc AP = r Φ ; thus OA = r Φ i
   [  Φ is angle PCA  &  θ is angle DCP ]

3. AC' = r j

4. C'P = r U_θ = r cos θ i + r sin θ j

5. Φ + θ = 3π/2    or  θ = 3π/2 - Φ ;

       in #4. C'P = r cos(3π/2 - Φ) i + r sin(3π/2 - Φ) j

6. from trig: cos(3π/2 - Φ) = cos 3π/2 cos Φ + sin 3π/2 sin Φ = - sin Φ

        sin(3π/2 - Φ) = sin 3π/2 cos Φ - cos 3π/2 sin Φ = - cos Φ

      so   C'P = - r sin Φ i - r cos Φ  j

7. thus               x = r Φ  - r sin Φ             the Parametric Equations

                    y = r Φ  - r cos Φ               of the Cycloid.


IV. Given a circle of radius "a". A circle of radius "b" rolls externally on

   the circumference of the fixed circle counterclockwise from the

   point A ( a, 0 ). Find the Parametric Equations of the Epicycloid generated.

    1. geometry: the straight line thru O & C passes thru the point of tangency T

    2. OP = x i + y j = OC + CP

    3. OP = [a + b] Uθ = [a + b] cos θ i + [a + b] sin θ  j

    [θ = angle AOT = angle GCE   and  α = angle TCP ]

    4. CP = b UΦ  = b cos Φ i + b sin Φ  j  where  Φ  =  θ + π  + α = π  + [θ + α ]

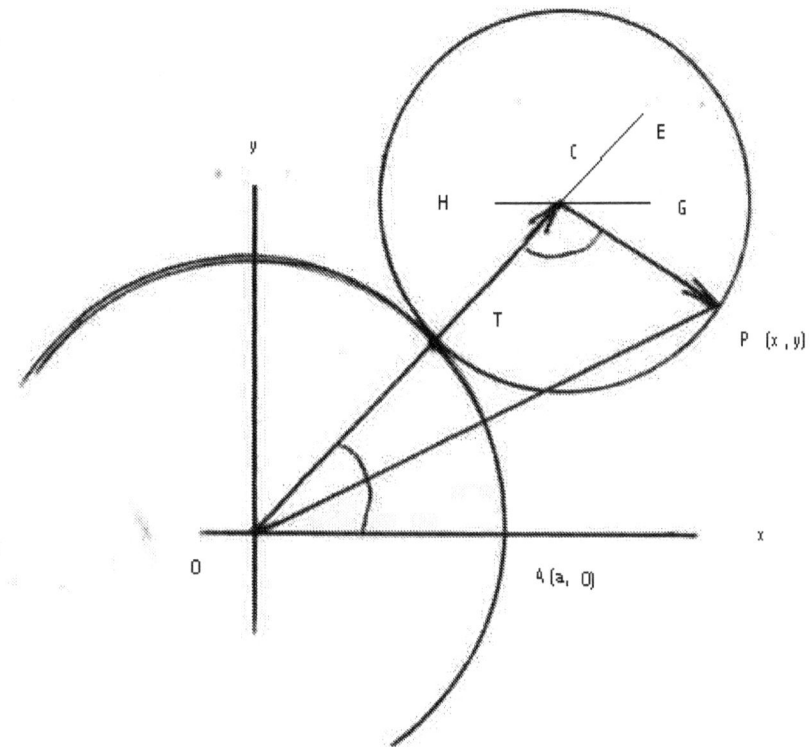

5. CP = b cos[ $\pi$ + ($\theta$ + $\alpha$)] i + b sin[ $\pi$ + ($\theta$ + $\alpha$)] j

   = -b cos ( $\theta$ + $\alpha$ ) i - b sin ($\theta$ + $\alpha$ ) j

6. roll the external circle back so that pt.P is on pt.A; then arc TP will lay on

arc TA. So, arc TP = arc TA

7. arc TA = a $\theta$    & arc TP = b $\alpha$ ; so a $\theta$ = b $\alpha$   or  $\alpha$ = a $\theta$ /b

8. into 5: CP -b cos[ (a + b) $\theta$ /b] i - b sin[ (a + b) $\theta$/b] j

9. 4 & 7 into 3:            x = (a + b) cos $\theta$ - b cos[ (a + b) $\theta$ /b]

                           y = (a + b) sin $\theta$ - b sin[ (a + b) $\theta$ /b]

5.0 Products of Two Vectors:

I. Introduction.

  A.1. While the definition of the vector sum of two vectors seems reasonable and obvious, a definition of an operation between  two vectors that might seem like a multiplication is not so obvious.

   2. There are three kinds of forms that have been discovered useful in physical applications that given the impression of multiplication.

 B. 1. DOT PRODUCT, A o B, which is a Scalar so is also called the SCALAR PRODUCT. From one of its forms seen later is it is reasonable to call it the INNER PRODUCT.

   2. CROSS PRODUCT, A x B, which is a Vector. It is also called the VECTOR PRODUCT and also the OUTER PRODUCT.

   3. DYADIC PRODUCT, A B, which is a Dyad. For a while Matrix Theory replaced a need for this product, but Data Processing found a  use for it again. We will not study this product.

II. The DOT PRODUCT.

  A.1. The following is one way to give meaning to a definition for the operation,

   2. Before we have tried to expand many of the operational concepts of the Real Number system to have meaning in a Vector Space; so to try to give meaning to the squaring of a binomial, say, consider that in form it would look like:

$[A - B]^2$  = $[A - B]$ * $[A - B]$ where at the moment it is clear that * is some kind of a multiplication but it is undefined because the elements are vectors not numbers.

   3. if * is a product that acted like numbers, then we could write:

$[A - B]^2$  = $[A - B]$ * $[A - B]$ = $A^2$ $-2A$ * $B + B^2$ ; if we decided to write it as:

$[A - B]^2$ = $A^2$ + $B^2$ - 2 A * B   we might note that the form looks familiar

   4. Consider A & B with $0 \leq \theta \leq \pi$   as the angle between them: from the figure on page 54 find A - B ; then by the Law of Cosines:

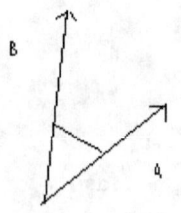

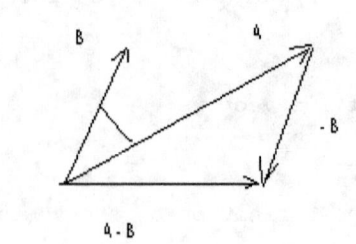

$|A - B|^2 = |A|^2 + |B|^2 - 2 |A| |B| \cos \theta$

Remember $|A|$ is the length of the vector A

and is a positive real number; so $[A - B]^2 = |A - B|^2$, and $A^2 = |A|^2$, and $B^2 = |B|^2$

5. from all of this it seems like $[A * B]$ behaves like $[|A| |B| \cos \theta]$,

6. This gives us a good idea for a definition for A * B, but we will change the *

to a special Dot, o , for the product of two vectors    [ some use • ].

B. DEFINITION of the DOT PRODUCT:

A o B = $|A| |B| \cos \theta$ where $\theta$ is the angle between the two vectors.

C. Because cos ( $-\theta$ ) = $\cos \theta$  and cos ($2\pi - \theta$) = $\cos \theta$ it is immaterial

how we describe the angle; so for convenience we restrict it to $0 \leq \theta \leq \pi$

We consider $\theta$ as the angle formed as A is rotated counterclockwise into B. Note we

are using the concept of Free Vectors and that the forms of A & B that have the same

initial point.

D. We see that the A o B operation between Vectors is a Real Number.

III. Note:

1. Given A & B , then A o B = $|A| |B| \cos \theta$

2. and B o A = $|B| |A| \cos ( -\theta) = |A| |B| \cos \theta$ = A o B

3. so  A o B = B o A  or The Dot Product of Vectors is Commutative.

IV. The DOT PRODUCT in terms of Components  [in 2-space]:

A. 1. let A = $\langle a_1 , a_2 \rangle$  and B = $\langle b_1 , b_2 \rangle$

2. find a third  vector C so that

C = B - A = $\langle b_1 - a_1 , b_2 - a_2 \rangle$ as in  $\Delta$ 1

Note: $|C| = \sqrt{ [(b_1 - a_1 )^2 + (b_2 - a_2 )^2 ]}$

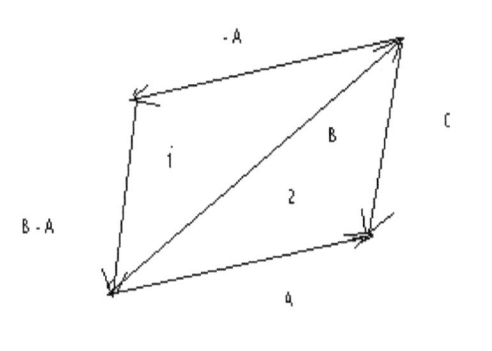

3. in  $\Delta\,2$   from Law of Cosines: $|C|^2 = |A|^2 + |B|^2 - 2\,|A|\,|B|\,\cos\theta$

4. or: $|A|\,|B|\,\cos\theta = \tfrac{1}{2}\,[\,|A| + |B| - |C|\,]$

5. but: $|A|\,|B|\,\cos\theta = A \circ B$

6. so: $A \circ B = \tfrac{1}{2}[(a_1^2 + a_2^2) + (b_1^2 + b_2^2) - (b_1 - a_1)^2 - (b_2 - a_2)^2]$

   $= \tfrac{1}{2}[a_1^2 + a_2^2 + b_1^2 + b_2^2 - b_1^2 + 2a_1\,b_1 - a_1^2 - b_2^2 + 2a_2\,b_2 - a_2^2]$

7. $A \circ B = \tfrac{1}{2}\,[2a_1\,b_1 + 2a_2\,b_2] = a_1\,b_1 + a_2\,b_2$ .

B. So we have proved:

   The Theorem: $A \circ B \equiv \langle a_1, a_2\rangle \circ \langle b_1, b_2\rangle = a_1\,b_1 + a_2\,b_2$ .

note: we could use this as the definition and prove the other as a theorem.

   C. Exercise: Given A $=\langle1,2\rangle$ & B $=\langle6,3\rangle$. Fine the angle, $\theta$ ,between them.

       1. from IV.B. $A \circ B = [1][6] + [-2][3] = 6 - 6 = 0$

       2. from II.B. $A \circ B = \sqrt{[1+4]}\;\sqrt{[36+9]}\;\cos\theta = \sqrt{5}\;\sqrt{45}\;\cos\theta$

           $A \circ B = \sqrt{225}\;\cos\theta = 15\cos\theta$

       3. so: $\cos\theta = 0/15 = 0$   or  $\theta = \pi/2$

   D. Given the function $P[A \circ B] = 3\,[A \circ B]$. Is P Linear?

       1. $P[(A \circ B) + (C \circ D)] = 3[(A \circ B) + (C \circ D)] = 3(A \circ B) + 3(C \circ D)$

                           $= P[A \circ B] + P[C \circ D]$,          checks

       2. $P[k\,(A \circ B)] = 3[k\,(A \circ B)] = k[\,3(A \circ B)]$

               $= k\,[P(A \circ B)]$,                    checks

       3. Yes this function of a Dot Product is Linear.

5.1 Basic Theorems: Given: A $= \langle a, a\rangle$, B $= \langle b, b\rangle$, C $= \langle c, c\rangle$, D $= \langle d, d\rangle$

   I.   Theorem :        $A \circ (B + C) = A \circ B + A \circ C$

1. $A \circ (B + C) = <a_1, a_2 > [ <b_1, b_2 > + <c_1, c_2 >]$

   $= <a_1, a_2 > \quad <(b_1 + c_1), (b_2 + c_2)>$     vector sum

   $= a_1 (b_1 + c_1) + a_2 (b_2 + c_2)$     from IV.B

   $= a_1 b_1 + a_1 c_1 + a_2 b_2 + a_2 c_2$     algebra

   $= (a_1 b_1 + a_2 b_2) + (a_1 c_1 + a_2 c_2)$     algebra

2. so $A \circ (B + C) = A \circ B + A \circ C$        IV.B

II. Theorem :    $(A + B) \circ C = A \circ C + B \circ C$

   1. $(A + B) \circ C = C \circ (A + B) = C \circ A + C \circ B$

   2. so: $(A + B) \circ C = A \circ C + B \circ C$

III. Theorem :    $(A + B) \circ (C + D) = A \circ C + A \circ D + B \circ C + B \circ D$

   1. $(A + B) \circ (C + D) = (A + B) \circ C + (A + B) \circ D$

                           $= A \circ C + B \circ C + A \circ D + B \circ D$

   2. so: $(A + B) \circ (C + D) = A \circ C + A \circ D + B \circ C + B \circ D$

One can remember this by the FOIL mnemonic:

work out the products of the FIRST, OUTER, INNER, then LAST terms. We note

that we still have the same relations as we had with the Real Numbers.

IV. Theorem :    If $A \perp B$, then $A \circ B = 0$.

   1. If $A \perp B$, then $\theta = \pi/2$ and $A \circ B = |A| |B| \cos \pi/2 = 0$

   2. so: If $A \perp B$, then $A \circ B = 0$.

V. Theorem :    If $A \circ B = 0$ where $A \neq 0$ & $B \neq 0$, then $A \perp B$.

   1. If $A \circ B = 0$, then $|A| |B| \cos \theta = 0$

   2. then either $|A| = 0$, Or $|B| = 0$; Or $\theta = \pi/2$ which implies $A \perp B$

   3. so If $A \circ B = 0$ where $A \neq 0$ & $B \neq 0$, then $A \perp B$.

VI. Theorem :    If $A \circ B < 0$, then $\pi/2 < \theta < \pi$ and Conversely.

   1. in $A \circ B = |A| |B| \cos \theta$ we know $|A| > 0$ & $|B| > 0$ [or $|A|, |B| = 0$]

   2. if $A \circ B < 0$, then $\cos \theta < 0$, Or then $\pi/2 < \theta < \pi$

   3. so: if $A \circ B < 0$, then $\pi/2 < \theta < \pi$ and Conversely.

VII. Theorem :    $A \circ A = |A|^2 > 0$.

   1. $A \circ A = |A| |A| \cos \theta = |A| |A| \cos \theta = |A| |A|$ [1]

   2. so: $A \circ A = |A|^2 > 0$.

   VIII. Theorem :                    (r A) o (s B) = r s (A o B)

           1. (r A)  o (s B) = |r A| |s B| cos θ = r s |A| |B| cos θ

           2. so:  (r A) o (s B) = r s (A o B)

 Note: The Dot Product is between two Vectors; so A o B o C would  have no meaning;

no matter what association is set up, eventually there would have to be a Dot

between a vector and a scalar.

IX. Exercises – Prove: 1.  A o ( b B + c C )  = b A o B  +  c A o C .

                   2.  [A o B] + C  = A o C  +  B o C


5.2. In 2.16 we had the PROJECTION of B onto A. Review this and look at some

theorems, examples, problems connected with this concept.

   A. As:

      1. Drop a perpendicular, ⊥ , from pt B to A  forming  THE  VECTOR PROJECTION

         of B onto A

      2. This is symbolized by    Proj $_A$ B  ; this is a Vector  || A

      3. Its LENGHT is called THE SCALAR PROJECTION of B onto A  ; so:

      The Scalar Projection = + | Proj $_A$ B | = + |B| cos θ

         where a minus sign is used when A & B are oppositely extended.

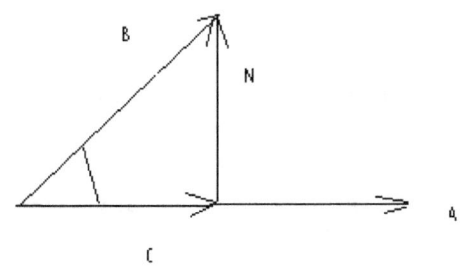

B.   1. A o B = |A| times the Scalar Projection of B onto A from step A.3.

     2. OR in a like manner we can show that:

      A o B = |B| times the  Scalar Projection of A onto B

C.  1. Make a Vector, N , out of the ⊥ to A from B but direct it toward B,

      2. let  Proj $_A$ B = C ; then  C = k A :k > 0

      3. N = - C + B = - k A + B = B - k A : k > 0

4. $A \circ N = |A| \ |N| \cos \pi/2 = 0$

5. #3 into #4: $A \circ (B - kA) = 0$ or $A \circ B - A \circ kA = 0$

6. or $A \circ B - k (A \circ A) = 0$

7.   $k = (A \circ B) / |A|^2$   which is a scalar

8.   so: $N = B - [ (A \circ B)/|A|^2 ] A$          #7 into #2

D.

1. from C.3. : $C = B - N = B - \{ B - [ (A \circ B)/|A|^2 ] A \}$

2. so: $\text{Proj}_A \ B = [ (A \circ B)/|A|^2 ] A$

E . Theorem;  Prove $N = \langle a, b \rangle \perp$ to the straight line $ax + by + c = 0$

            Where not both a & b can be zero.

1. Take two distinct points $P_1 (x_1, y_1)$ & $P_2 (x_2, y_2)$

   on the line $ax + by + c = 0$

2.   $a x_1 + b y_1 + c = 0$

     $a x_2 + b y_2 + c = 0$

3.  $a [x_1 - x_2] + b[y_1 - y_2] = 0$

4. now $P_1 P_2 = [x_1 - x_2] i + [y_1 - y_2] j$

5. so $N \circ P_1 P_2 = \langle a, b \rangle \circ [x_1 - x_2] i + [y_1 - y_2] j = a[x_1 - x_2] + b[y_1 - y_2]$

            $= 0$      from #3

6. If $N = 0$, then $\langle a, b \rangle = 0$ but only true if both $a = 0$ & $b = 0$; given not both can be 0; so $a \neq 0$ & $b \neq 0$ --- thus $N \neq 0$

7. If $P_1 P_2 = 0$ then pt. $P_1$ coincides with pt. $P_2$, but taken distinct; so $P_1 \circ P_2 \neq 0$

8. now $N \circ P_1 P_2 = |N| \ |P_1 \circ P_2| \cos \theta = 0$  iff  $\theta = \pi/2$

9. thus $N \perp P_1 P_2$

F. now  $A \circ A = |A|^2$ ; so $|A| = \sqrt{[A \circ A]}$

G. Find the distance, d, from point P (4,3) to line l: $x + 3y - 6 = 0$

1. take pt. B (0,2) as a convenient point L  [ note: $0 + [3][2] - 6 = 0$ ]

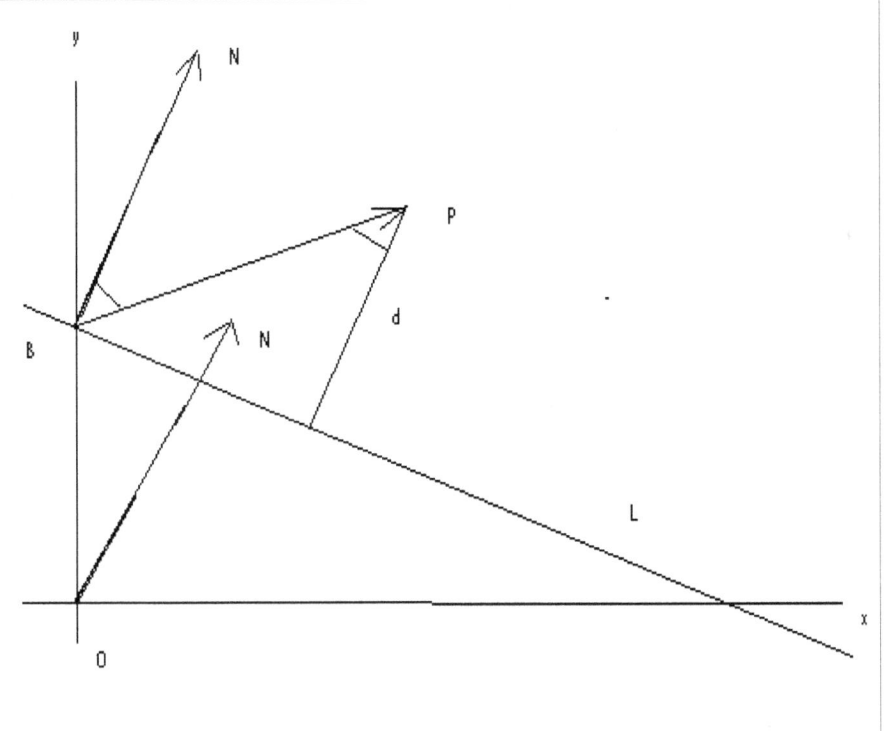

2. draw N thru pt. B ⊥ l , then from the theorem in E.  :

   N = <1,3>; thus |N| =  √[1+9] =  √10

3. d = |BP| cos θ    but N o BP = |N| |BP| cos θ = |N| d

   d = [N o BP]/|N|

4. BP = <4-0,3-2> = <4,1>

5. so d = [<1,3> o <4,1>]/  √10 = [(1)(4) + (3)(1)]/  √10  = [4+3]/  √10

      d = 7/ √10

H. Apply a constant force F on a pt. A to get pt. A to move to pt. B along the
line AB. Find the Work W done on pt.A  [Remember Work is Force times Distance].

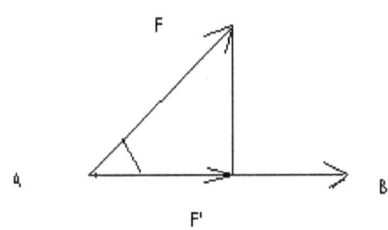

1. The amount of the Force, F', that acts along the line AB

   is  F' = |F| cos $\theta$

2. W = [ |F| cos $\theta$ ] [ |AB| ] = |F| |AB| cos $\theta$

3. W =  F o AB.

I. now  A o B = |A|  |B| cos $\theta$

So    cos $\theta$ = [A o B]/ {|A||B|}

The angle, $\theta$ , between two vectors :  $\theta$  =  cos$^{-1}$ [ A o B  / { |A|  |B| }]

J. Because the Dot Product of two vectors is a Positive, Real Number

We see that:   | A o B |  =  A o B


5.3 Prove: The Cauchy-Schwarz Inequality: | A o B | $\leq$ |A| |B|

Case I. Assume A = 0

1. so |A|  =0

2. now |A o B| =  |A|   |B|  cos $\theta$ = (0)  |B|  cos $\theta$ = 0 = (0) |B|  =  |A|  |B|

3.     |A o B|  = |A| |B|       the = part checks for   0 $\leq$ $\theta$ $\leq$ $\pi$

   Note the proof and the results will be the same if  B = 0;

Case II. Assume A || B where A $\neq$ 0, B $\neq$ 0; implies  A = k B or B = k A

      1. in either case    $\theta$ = 0 and cos 0 = 1

      2. A o B  =  |A|  |B|  cos $\theta$ =  |A|  |B|  (1) > 0

      3. A  o B  =  |A|  |B|

      4. Because A o B is a real, positive number this can be written as

         | A o B |  =  |A|  |B|       checks for  0 $\leq$ $\theta$ $\leq$ $\pi$

Case III. Assume A $\neq$ 0  &  B $\neq$ 0   &  0 $\leq$ $\theta$ $\leq$ $\pi$/2  [ note: A || B ]

      1. from $\theta$ :  0 < cos $\theta$ $\leq$ 1

      2. |A|  > 0 &  |B|  > 0 ; so |A|    |B| > 0

      3. Multiplying #1 by #2 : |A|  |B|  0 < |A|  |B| cos $\theta$ < |A|  |B| $\cdot$ 1

      4. so 0 < A  B  <  |A|  |B|

      5. cos $\theta$ > 0; so  |A|  |B|  cos $\theta$ > 0   &   A  B > 0

6.  | A ∘ B | = A  ∘ B    both give the same positive {or 0}, real number

7.  #6 into #4: | A  ∘ B } <  |A|  |B|    for 0 < $\theta$ ≤ $\pi/2$            checks

Case IV.  A ≠ 0 & B ≠ O &   $\pi/2 < \theta < \pi$

1.  −1 < cos $\theta$ < 0

2.  |A|  |B|  (−1) <  |A|  |B|  cos $\theta$ <  |A|  |B| · 0

3.  − |A|  |B|   < A ∘ B < 0

4.  multiplying by (−1): |A|  |B|  > − A  B > 0 or − A  B < |A|  |B|

5.  cos $\theta$ < 0 ; so A ∘ B < 0 or − A  B > 0

6.  − A ∘ B = | A  ∘ B | both give the same positive real number

7.  #6 into #4: |A  ∘ B | <  |A|  |B| for   $\pi/2 < \theta < \pi$            checks

Conclusion. | A ∘ B | < |A||B|     for  0 ≤ $\theta$ ≤ $\pi$   and for all A's & B's


5.4  Exercise. Show the inequality true by using A = <$a_1$  ,$a_2$ > &   B= <$b_1$  ,$b_2$ >.


5.5. Note:

1. The proof in 5.3 is a "Constructive Proof"; it starts with the givens then develops the material of the proof step by sequence step to the "to prove" of the theorem.

2. The following is a "Verification Proof":

1. the |A ∘ B| < |A||B| can be written as: | |A||B| cos $\theta$ | < |A||B|

2. or |A| |B| |cos $\theta$ | < |A||B|

3. Magnitudes of vectors are Positive, Real Numbers;

    so divide by |A||B|, to get |cos $\theta$ | < 1 which implies

    −1 < cos $\theta$ < 1; a true Trigonometry relation.

4. so  the thereom is verified.


5.6 Prove THE TRIANGLE INEQUALITY : |A  + B | ≤ |A| + |B|

    where the Equality occurs iff either A = 0 or B = 0

    or A = c B: c>0 [this is  the same thing as A || B with same direction]

Case 1. Prove: $|A + B| = |A| + |B|$

    1. If $A = 0$ , then $|A| = 0$ ;

       and $|A + B| = |0 + B| = |B| = 0 + |B| = |A| + |B|$

    2. If $B = 0$ , then same conclusion with same proof

    3. If $A = c B$: $c>0$,

      then $|A| = |c B| = c |B|$,

      now $|A + B| = |c B + B| = |(c + 1) B|$

             $= (c + 1) |B|$  for $c>0$ implies $c + 1 > 0$

             $= c |B| + 1 \cdot |B| = c |B| + |B|$

             $= |A| + |B|$

Case 2. Prove $|A + B| < |A| + |B|$ where $A \neq 0$, $B \neq 0$, & $A$ is not $||$ $B$.

    1. we have seen $|A + B|^2 = [A + B] \circ [A + B] = A \circ A + A \circ B + B \circ A + B \circ B$

                   $= |A|^2 + 2 [A \circ B] + |B|^2$

    2. from the Cauchy-Schwarz Inequality:  $A \circ B < |A| |B|$

  3. so from 1.   $|A + B|^2 < |A|^2 + 2 |A||B| + |B|^2$

                $< [ |A| + |B| ]^2$

4. taking the positive Square Root of both sides:

        $|A + B| < |A| + |B|$

5.Conclusion: the Theorem is proved.


5.7 Notes:

    I. Forming a triangle from $A$ & $B$ & $[A + B]$,

      from geometry we note that

      $|A| + |B| > |A + B|$ as was true in the inequality.

    II. Take $A = <a_1 , a_2 >$ & $B = <b_1 , b_2 >$ to show

      the two theorems in component form:

The Cauchy-Schwarz Inequality:    $|a_1 b_1 + a_2 b_2 | \leq \sqrt{[a_1^2 + a_2^2]} \cdot \sqrt{[b_1^2 + b_2^2]}$

The Triangular Inequality: $\sqrt{\{[a_1 + b_1 ]^2 + [a_2 + b_2 ]^2\}} \leq \sqrt{[a_1^2 + a_2^2]} \cdot \sqrt{[b_1^2 + b_2^2]}$

5.8 Theorem -- a modified "triangle inequality":

1. from 5.6: |A| + |B| ≥ |A + B|: in words this says "the sum of the absolute values [or norms] of  two vectors is greater than or equal to the absolute value of the sum of the two vectors". So,

$$[ \ |A - B| \ ] + |B| \geq | \ [A - B] + B \ |$$

$$\geq | \ A - B + B \ |$$

$$\geq | \ A \ |$$

2. Subtracting  |B| > 0 from both sides:

|A - B| ≥ |A| - |B|          the modified Triangle Inequality.

6.0 The Cartesian Coordinate System in 3-Space:

The "real world" exists in three dimensions ; so we will expand our previous

concepts to 3-space.

The student is familiar with the 3-space coordinate system shown in the figure

with the equivalent/equal class of vector CD. We see that CD = OA which means that

we can also write CD as just A. This will be  the graphic basis of all vectors in

3-space.

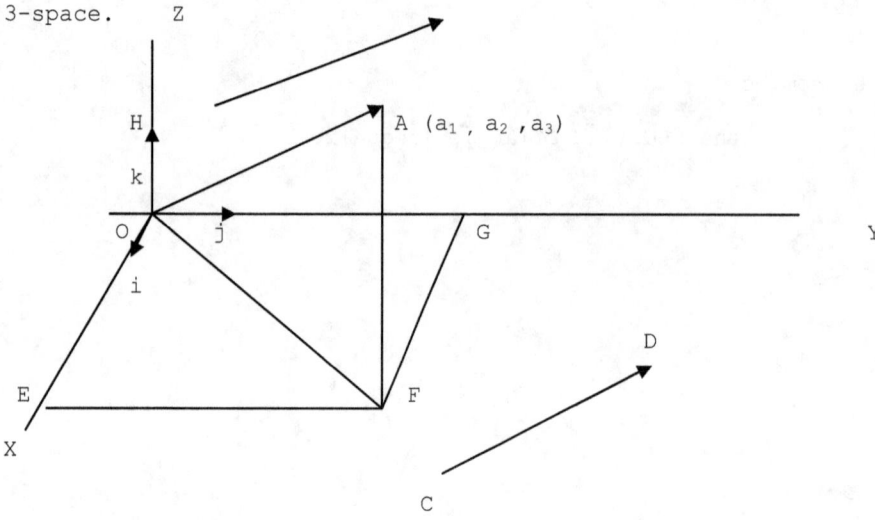

6.1

   1. The Position Vector, A, will still be defined as that vector whose tail is on

the origin and whose head is the point A ($a_1$ ,$a_2$ ,$a_3$ ) ,

   2. There is a 1-to-1 correspondence between the point A ($a_1$ ,$a_2$ ,$a_3$ ) and the

vector OA ,

   3. So, The Definition of The POSITION VECTOR is: OA = A = <$a_1$  ,$a_2$  ,$a_3$  >.

6.2

 1. To get started in 3-space we will use geometry. While Solid Geometry is not as

well known as plane geometry and is difficult to represent graphically, its

solutions in many cases are simpler than those of the algebraic developments.

 2. A study of the figures in the next sections will make clear the reasons why many

of the identities listed are true,

3. We will find that in many cases the study here just involves adding on another component to the items of the previous study. Many of the definitions and theorems do not even have to be revised to expand them into 3-space,

4.  By Definition a SCALAR PRODUCT is the Vector c A where c A || A; and |c A| = |c| |A|; also if c>0, then c A has the same direction as A; and if c<0, then c A has the opposite direction as A --- no matter the dimension of A.

As:    $cA = c<x, y, z> = <cx, cy, cz> \equiv cx\ i + cy\ j + cz\ k$

6.3  Using the same kind of reasoning that we used in 2-space to give meaning to our definitions, we have the following in 3-space:

1. Find the algebraic sum of  vectors A $<a, b, c>$ & B $<d, e, f>$ :

$$A + B = <a,b,c> + <d,e,f> = <a+d,\ b+e,\ c+f>$$

2. Given $A(a_1, a_2, a_3)$ & $B(b_1, b_2, b_3)$ two points in 3-space, find the vector AB:

a. draw OA & OB

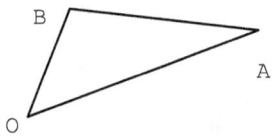

b. $AB = AO + OB = -OA + OB = <b_1, b_2, b_3> - <a_1, a_2, a_3>$

c. $AB = < b_1 - a_1,\ b_2 - a_2,\ b_3 - a_3 >$

   note: if we let $b_1 - a_1 = c_1$ & $b_2 - a_2 = c_2$ & $b_3 - a_3 = c_3$,
   then $<c_1, c_2, c_3>$ || AB.

6.4 from Figure 6.0 find the Magnitude , |A| , of the vector A:

1. $A = <a_1, a_2, a_3> = a_1\ i + a_2\ j + a_3\ k$

   where $i = <1,0,0>$ , $j = <0,1,0>$ , $k = <0,0,1>$

2. $|OF|^2 = a_1^2 + a_2^2$    so  $|OA|^2 \equiv |A|^2 = |OF|^2 + a_3^2$

$$|A| = \sqrt{[a_1^2 + a_2^2 + a_3^2]}$$

6.5 Dot [or Scalar] Product where $A = <a_1, a_2, a_3>$ & $B = <b_1, b_2, b_3>$.

I. 1. We still have by Definition: $A \circ B = |A|\ |B|\ \cos\theta : 0 \leq \theta \leq \pi$,

   where $\theta$ is the angle from A to B.

2. EXERCISE: see 5.0-IV. and prove $A \circ B = a_1 b_1 + a_2 b_2 + a_3 b_3$

3. If $A \perp B$ , then $\theta = \pi/2$ and $A \circ B = 0$ and conversely.


## 6.6. Directional Cosines from figure 6.0:

let angle EOA = $\alpha$ , angle COA = $\beta$ , & angle AOH = $\gamma$

1. $OE \circ OA = <a_1, 0, 0> \circ <a_1, a_2, a_3> = a_1^2 + 0 + 0 = a_1^2$

and $OC \circ OA = a \cdot \sqrt{[a_1^2 + a_2^2 + a_3^2]} \cdot \cos \alpha$ both from 6.5

so: $a_1 \cdot \sqrt{[a_1^2 + a_2^2 + a_3^2]} \cdot \cos \alpha = a_1^2$

2. in the same manner:

$a_2 \cdot \sqrt{[a_1^2 + a_2^2 + a_3^2]} \cdot \cos \beta = a_2^2$

$a_3 \cdot \sqrt{[a_1^2 + a_2^2 + a_3^2]} \cdot \cos \gamma = a_3^2$

3. thus using OC: $\cos \alpha = a_1 / \sqrt{[a_1^2 + a_2^2 + a_3^2]} = a_1 / |A|$

and using OD: $\cos \beta = a_2 / \sqrt{[a_1^2 + a_2^2 + a_3^2]} = a_2 / |A|$

and using OE: $\cos \gamma = a_3 / \sqrt{[a_1^2 + a_2^2 + a_3^2]} = a_3 / |A|$

4. These are called The DIRECTION COSINES of A;

$\alpha, \beta, \gamma$ are The DIRECTION ANGLES of A.


## 6.7 INEQUALITIES that are still true for any Vectors A = <a, b, c> & B = <d, e, f>:

I.

1. THE CAUCHY-SCHWARZ INEQUALITY: $|A \circ B| \leq |A| |B|$;

or $|a d + b e + c f| \leq \sqrt{[a^2 + b^2 + c^2]} \cdot \sqrt{[d^2 + e^2 + f^2]}$

2. THE TRIANGLE INEQUALITY: $|A + B| \leq |A| + |B|$

or $\sqrt{[[a + d]^2 + [b + e]^2 + [c + f]^2]} \leq \sqrt{[a^2 + b^2 + c^2]} + \sqrt{[d^2 + e^2 + f^2]}$

II. If the vectors are written with subscripts as in $A = <a_1, a_2, a_3>$

and $B = <b_1, b_2, b_3>$, then the Inequalities can also be written as:

1. THE CAUCHY-SCHWARZ INEQUALITY; $|\sum a_k b_k| < [\sum a_k^2]^{1/2} [\sum b_k^2]^{1/2}$

2. THE TRIANGLE INEQUALITY: $[\sum (a_k + b_k)^2]^{1/2} < [\sum a_k^2]^{1/2} + [\sum b_k^2]^{1/2}$

In both cases $k = 1, 2, 3$.

6.8   We will now study some 3-space concepts that do not have 2-space counterparts.
We will start with another form that gives the impression of a product that occurs
in many studies of the real world - like work in the fields of Electricity and
Magnetism.

   We will jump back and forth between geometric and algebraic forms as we develop
the concepts.

   We will need some more than routine theorems and methods of proof of some
theorems from a course in Solid Geometry --- however nothing more exotic than what
would be covered in a high school course. Most of this material will be outlined by
graph and text because of a weakness that occurs in some courses of study of
geometry. Lines and vectors will be used interchangeably - whichever is most
convenient.

6.9   The Cross  [ or Vector ] Product.

  I.   1. Given two vectors A & B,

      2. Choose their equals that have their initial points at point O, and let  θ
be the angle between them where  $0 \leq \theta \leq \pi$

      3. A & B  determine a plane P in 3-space      A x B

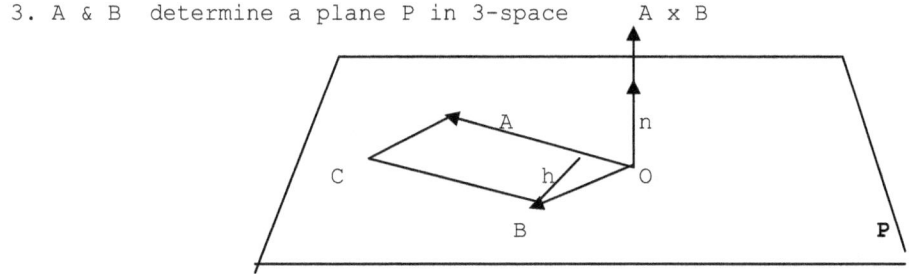

      4. Let n be a Unit Vector ⊥ plane P  that points in the direction a
right-handed screw would advance when its Head is rotated from A to B by
angle θ. This sets up a RIGHT-HAND SYSTEM with A, B, n .

  II. Definition: The VECTOR PRODUCT or The CROSS PRODUCT of A & B is the VECTOR:

      A x B = n • |A||B| sin θ  : $0 < \theta < \pi$

      Note: n ⊥ A , n ⊥  B , n ⊥ the plane P of A & B.

III.

    1. If A || B , there is still a plane P and $\theta$ = 0 or $\theta$ = $\pi$

so sin $\theta$ = 0; thus if A || B , then  A x B = 0

    2. If either A = 0 or B = 0 , then A x B = 0.

6.10 Theorems:  see figure 6.9

  I. In the Plane of A & B:

    1. sin $\theta$ = h / |B|      so h = |B| sin $\theta$

    2. and |A| |B| sin $\theta$  is the Area of the

       parallelogram OACB,

    3. Thus,  Area of OACB = | A x B |

  II.

    1. If we look at B x A , we note that its n vector is the

       Negative of A x B 's n,

    2. so B x A = -n |A| |B| sin $\theta$,

    3. or    B x A = - A x B : The Cross Product is Not Commutative;

                                    is Anti-Commutative.

  III.

    1. the n of [i x j] is k; so  i x j = k |i||j| sin $\pi$/2

                            = k $\cdot$ 1 $\cdot$ 1 $\cdot$ 1  = k

    2. in the same fashion; note the mnemonic graphic device:

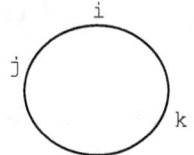

                                      Cross any Two get the Third;

                                      Counterclockwise + ; Clockwise -

      i x j = - j x i =  k

      j x k = - k x j = i

      k x i = - i x k = j

    3. if $\theta$ = 0  then  sin $\theta$ = 0 ; so  i x i = j x j = k x k = 0

      in general   A x A = 0

IV.

   1. r A is a vector that lies along the same line as the vector A.

      s B is a vector that lies along the same line as the vector B. If θ is the

angle between A & B, then it is also the angle between r A & s B.

   2. [ r A ] x [ s B ] = n |r A| |s B| sin θ = rs [ n |A||B| sin θ ]

   3.  so    [ r A ] x [ s B ] = rs [A x B];

   4.  or       The Cross Product [of 2 vectors] is Associative with Scalars.

V.  Note:

   1. i x [j x k] = i x i = 0

      [i x j] x k = k x k = 0

   2. so  i x [j x k] = [i x j] x k  ; or this kind of  Cross Product of

      the i, j, k Vectors is also Associative.

   3. but:    while        j x [j x k] = j x i  = - k,

              we have      [j x j] x k  = 0 x k = 0;    not the same; so

         this kind of Cross Product of the i, j, k   Vectors is not associative.


6.11 Geometric Construction of A x B;   A,B ε V  &  A not parallel to B.

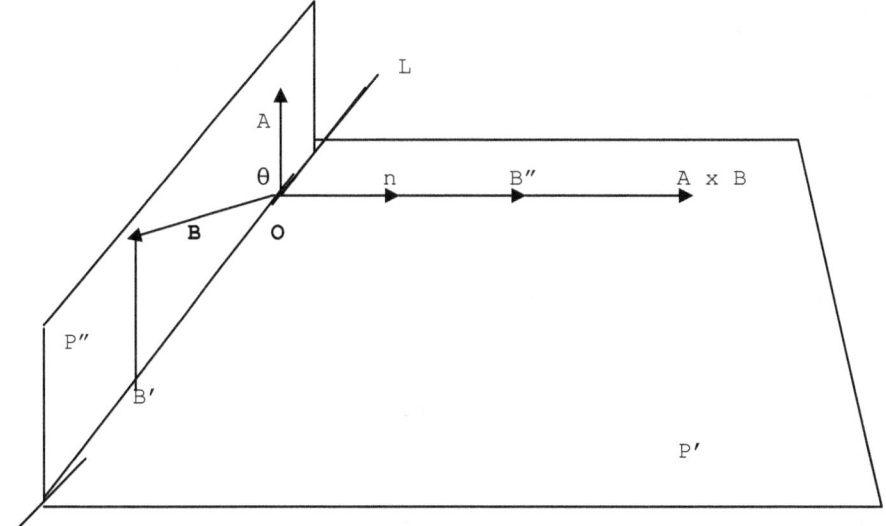

1. A & B determine a plane P″ ; take the initial points of A & B at the origin, O,

2. both are Position Vectors and θ is the angle between them where $0 < θ < π$

3. Construct a Plane P′ ⊥ A at point O

4. By a geometry theorem any plane through A is ⊥ to P′ ; so P′ ⊥ P″

5. P′ & P″ intersect in the Straight Line L through Pt. O ; so A ⊥ L

6. from every point of B drop perpendiculars to P′ [ or Orthogonally Project B onto P′ ]; this gives us B' the projected vector to plane P′ which is along L and through Point O

7. the projection, B B', is ⊥ L; thus the triangle is a Right Triangle with the right angle at B' as indicated in the figure,

8. B B' || A and angle B'BO = θ

9. sin θ = |B'| / |B| or |B'| = |B| sin θ

10. in P′ rotate B' by 90° about O to form B''; so |B''|=|B'|

    or |B''| = |B| sin θ

11. B" ⊥ A [if a line is ⊥ to a plane then it is ⊥ to all lines in the plane through the foot of the ⊥ ] and B" ⊥ B,

12. B" ⊥ P″ [a line ⊥ to two intersecting lines in a plane is ⊥ to the plane]

13. form the Unit Vector n = B"/ |B"| ; n ⊥ the plane, P″ , of A & B , and it fits a Right-Hand system

14. n = B"/ |B"| = B"/{|B| sin θ }= |A| B" / {|A| |B| sin θ} ;

    so |A| B" = n |A| |B| sin θ = A x B

15. the vector A x B is the vector in P′ obtained by finding the scalar product of the scalar |A| and the vector B″

II. In words to get the Cross Product between a Vector A and a second vector, B, we project the B onto the plane, P′, that is ⊥ A and then rotate that projected vector by +90° in P′ to get A x B when the rotated vector is multiplied by |A|.

6.12 Component form of the Cross Product.

Given:  $A = <a_1, a_2, a_3>$  and  $B = <b_1, b_2, b_3>$

A is not parallel to B  [ $\theta \neq 0$ or $\theta \neq \pi$ ]

Find:  A x B

Note:  If we attempt to grind this out, we find that we need a Distributive Law which we will take up later. Dot Products involve $\theta$ and perpendiculars, and a trick with Dots will give us what we want.

1. $A \perp [A \times B]$ ; so  $A \circ [A \times B] = |A|\ |A \times B| \cos \pi/2 = 0$

   $B \perp [A \times B]$ ; so  $B \circ [A \times B] = 0$

2. let $A \times B = <x, y, z>$ then

   $<a_1, a_2, a_3> \circ <x, y, z> = 0$  and  $<b_1, b_2, b_3> \circ <x, y, z> = 0$

3. thus  $\begin{cases} a_1 x + a_2 y + a_3 z = 0 \\ b_1 x + b_2 y + b_3 z = 0 \end{cases}$  $\begin{matrix} \alpha \\ \beta \end{matrix}$

4. Exercise: By mult $\alpha$ by $b_3$, $\beta$ by $-a_3$, & adding ; and so on ..... prove:

   $x = a_2 b_3 - a_3 b_2$ ; $y = a_3 b_1 - a_1 b_3$ ; $z = a_1 b_2 - a_2 b_1$

5. therefore  $A \times B = [a_2 b_3 - a_3 b_2]\ i + [a_3 b_1 - a_1 b_3]\ j + [a_1 b_2 - a_2 b_1]\ k$

6. also from previous algebra we see we can write this as

$$A \times B = \begin{vmatrix} i & j & k \\ a_1 & a_2 & a_3 \\ b_1 & b_2 & b_3 \end{vmatrix}$$

[because evaluating this 3<sup>rd</sup> order Determinant by expanding by the Co-Factors of the 1<sup>st</sup> Row we get step #5]. We will say much more about determinants in another chapter.


6.13  The Distributive Law for Cross Products:

Given vectors A, B, C were none are parallel, then $A \times (B + C) = A \times B + A \times C$

I.  The geometric proof of this theorem is a good example of the use of the theorems of Solid Geometry but complicated ; so we will prove it geometrically and algebraically.

II. Given vectors A, B, C were none are parallel. We have the following figures

for the geometric proof:

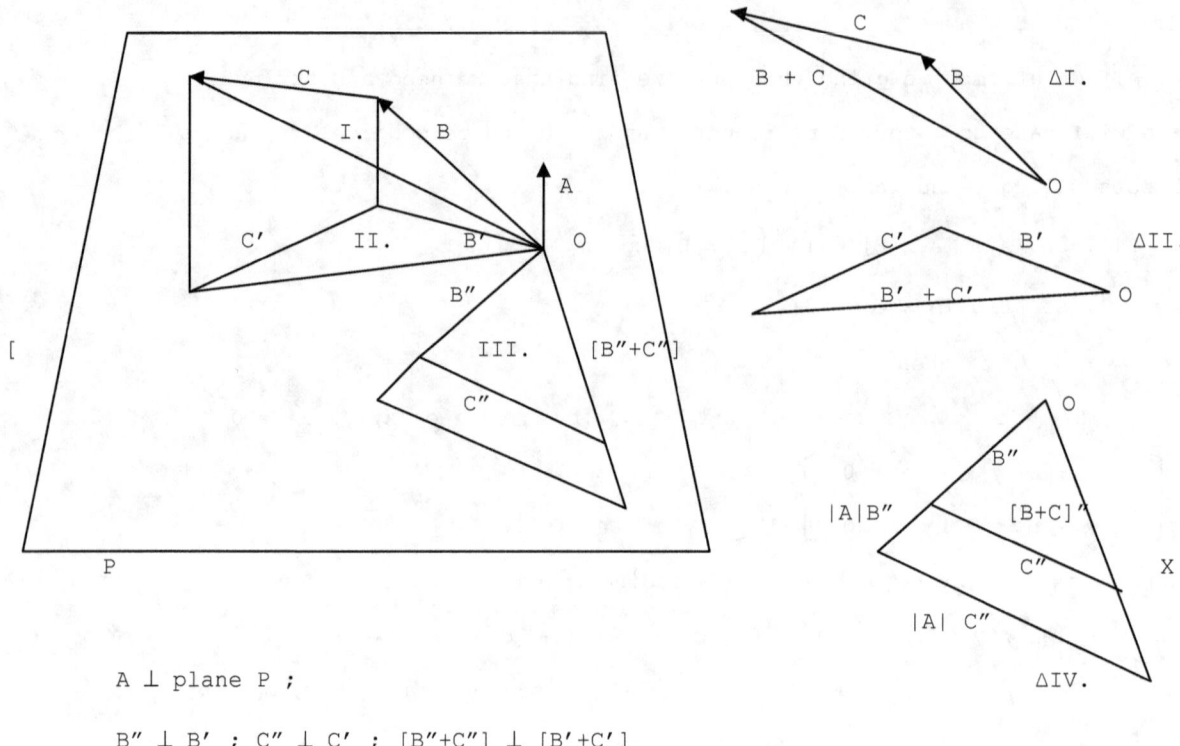

A ⊥ plane P ;

B″ ⊥ B′ ; C″ ⊥ C′ ; [B″+C″] ⊥ [B′+C′]

1. take A as a position vector and construct a plane P ⊥ A at point O,

2. take the B that is a position vector and add to it the vector C ; this forms a

triangle [ ΔI.] whose 3rd side is [B+C],

3. project this triangle on to P to form ΔII. whose sides are B′, C′, & [B′+C′];

 note: [B′+ C′] = [B + C]′  ,

4. rotate B′ by 90° in P to form ΔIII. with sides B″, C″, & [B″+C″]; note:

 [B″ + C″] = [B′ + C′]′ = [B + C]″. All sides will be rotated by 90° because:

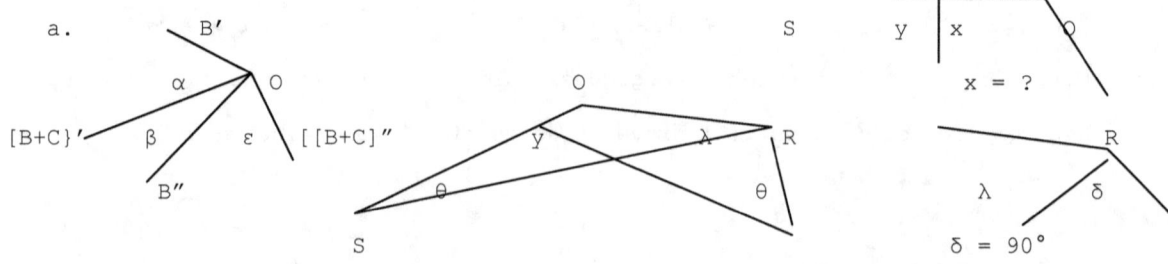

b. B″ ⊥ B′ ; so α + β = 90°

c. α = ε ; so ε + β = 90°  ; thus    [B″+C″] ⊥ [B′+C′]

d. angle O + θ + λ = 180°

    x + angle O + [λ + δ] + θ = 360°

    x + [angle O + θ + λ] + δ = 360°

    x + δ = 180°

    x = 90°  ; so  C″ ⊥ C′

5. expand ΔIII  by multiplying B″ & C″ by |A| to get side |A| B″ & vector |A| C″; draw |A| C″ as indicated || to C″ ; extend [B+C] as indicated to get ΔIV. with the 3ʳᵈ side X at the moment unknown:

a. because of Vector theory:  |A| C″ || C″

b. from geometry: | |A| B″ | / |B″| = X / | [B+C]″  or X = |A| [B+C]″

c. |A| [B″+C″], |A| B″ , & |A| C″ are the three sides of ΔIV.

6.therefore from vectors : |A| B″ + |A| C″ = |A| [B″+C″]

7. From 6.9 each of these is the equivalent Cross Product; so

A x B + A x C = A x ( B + C )  or   A x ( B + C ) = A x B + A x C

III. For the Algebra: take: A = $\langle a_1, a_2, a_3 \rangle$, B = $\langle b_1, b_2, b_3 \rangle$, C = $\langle c_1, c_2, c_3 \rangle$

1. B + C = $\langle b_1 + c_1, b_2 + c_2, b_3 + c_3 \rangle$

2. from 6.11 and a previous determinant theorem  [see again later]:

$$A \times [B + C] = \begin{vmatrix} i & j & k \\ a_1 & a_2 & a_3 \\ b_1 + c_1 & b_2 + c_2 & b_3 + c_3 \end{vmatrix}$$

$$A \times [B + C] = \begin{vmatrix} i & j & k \\ a_1 & a_2 & a_3 \\ b_1 & b_2 & b_3 \end{vmatrix} + \begin{vmatrix} i & j & k \\ a_1 & a_2 & a_3 \\ c_1 & c_2 & c_3 \end{vmatrix}$$

3. or  A x [B + C]  = A x B  + A x C

1. Note by use of the associative law we can write A x [ B + C + D ]

   as A x [ ( B + C ) + D and use 6.13. III. 3. to finally get :

   A x [ B + C + D ] = A x B + A x C + A x D

   And so on for larger sums.

2. We note: - [ A x [ B + C ] = - A x B - A x C

   or - ( - [ B + C ] x A ) = - ( - B x A ) - ( - C x A )

   and [ B + C ] x A = B x A + C x A

        which is a different form of the Distributive Law

3. From the previous work we see that :

   [ A x B ] ⊥ A      and [ A x B ] ⊥ B


6.15  Examples:

I.  Find the Area of the triangle ABC with A (-2, 3, 1) , B (1, -1, -2), & C (0, -2, 1)

   1. Find a point D so that a parallelogram ABCD is formed from the triangle;

   2. Area of ABCD = | AB x AC |

   3. AB = <1+2, -1-3, -2-1> = < 3, -4, -3 >

      AC = <0+2, -2-3, 1-1> = <2, -5, 0>

4.
$$\text{Area of ABCD} = \begin{vmatrix} i & j & k \\ 3 & -4 & -3 \\ 2 & -5 & 0 \end{vmatrix} = | < i [0-15] - j [-6-0] + k [-15+8] > |$$

        = | <-15 i + 6 j -7 k> | = √[225+36+49] = √310

5. Area of ABC = [√310] / 2


II. Find a Unit Vector, n , ⊥ to both A = <-2, -1, 2> and B = <3, -1, 1>

1. A x B is ⊥ to both A & B  & n = [A x B] / [|A x B|]

2.
$$A \times B = \begin{vmatrix} i & j & k \\ -2 & -1 & 2 \\ 3 & -1 & 1 \end{vmatrix} = <-1+2, 6+2, 2+3> = <1, 8, 5>$$

3. $| A \times B | = \sqrt{[1+64+25]} = \sqrt{90} = 3\sqrt{10}$

4. $n = \langle 1/3\sqrt{10}, 8/3\sqrt{10}, 5/3\sqrt{10} \rangle$

5. Exercise: check to see if $|n| = 1$.


## 6.16 Equation of a Plane.

I. Theorem.      Given: a point A [a,b,c] and vector $n = \langle d,e,f \rangle$

Find : the equation of the plane P that contains A and is $\perp$ n

1. Take any point R [x, y, z] in P

2. $AR = \langle x - a, y - b, z - c \rangle$

3. if $n \perp P$, then $n \perp AR$ ; so $n \circ AR = 0$

4. $\langle d, e, f \rangle \circ \langle x-a, y-b, z-c \rangle = 0$

   $d[x-a] + e[y-b] + f[z-c] = 0$

   $dx - da + ey - eb + fz - fc = 0$

   $dx + ey + fz = da + eb + fc$

5. a,b,c,d,e,f are given constants ; so let $da + eb + fc = k$   ,a constant

6. Equation of the plane P:   $dx + ey + fz = k$   where $n = \langle d,e,f \rangle \perp P$

   or                      $d(x - a) + e(y - b) + f(z - c) = 0$


II. Example: Find the equation of a plane P containing pt A [2,1,3] and

      $\perp$ to $n = \langle 3,-4,1 \rangle$

1. P:   $3x - 4y + z = k$

2. pt A is on P ; so $3[2] - 4[1] + [3] = k$ or $k = 5$

3 P:   $3x - 4y + z = 5$


III.  Exercise: Find the equation of the plane P that is $\perp$ to the two intersecting

planes R: $x - y + z = 0$ and Q: $2x + y - 4z = 5$ and contains the point A (4,0,-2)

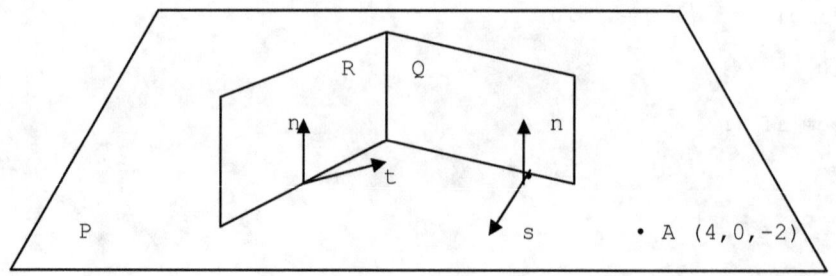

1. we see that from 6.16-I. the plane P that we are trying to find has the form of

d x + e y + f z = k; we have to find  the n ⊥ P & the scalars d,e,f,& k .

2. given: pl. P ⊥ pl. Q    and   pl. P ⊥ pl. R

3.                  s = <2,1,-4>  ⊥  pl Q

                 t = <1,-1,1>  ⊥  pl R

3. use the s that is on the line of intersection of planes Q & P; and
use the t that is on the line of intersection of planes R & P: they are Free Vectors,

4. s is in pl. P  and t is in pl. P [If two planes are ⊥  , a line {vector} ⊥ to of

them from any point of the other lies in the other plane]

5. take an n = <d,e,f> at the foot of s and an equal n = <d,e,f> at the foot of t

6. n ⊥ s   ; so <d,e,f>  <2,1,-4> = 0  or     2 d + e – 4 f = 0

   n ⊥ t   ; so <d,e,f>  <1,-1,1> = 0  or        d – e +  f = 0

7. adding the two equations: 3 d  – 3 f = 0   or  d =   f

   into the 2$^{nd}$ equation                      e = 2 f

8. P now looks like:  f x + 2 f y + f z = k

9. pt. A , (4, 0, -2 ) , is on P ; so  4 f + 0 – 2 f = k   or   k = 2 f

10. into step 8:   Plane P :  x + 2 y + z = 2 .


6.17.  A. If two planes intersect, they form a DIHEDRAL ANGLE. The angle θ between
The angle, θ ,  between the planes is the Angle between the two Normal Vectors to the
planes.

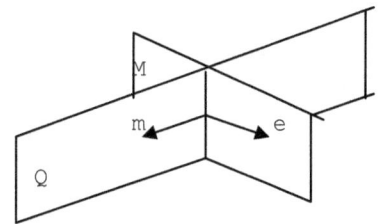

B. Examples:

I. Find the angle, θ , between pl.M, x-3y-z = 2,&  pl. Q,  2x-y-z = 3.

   1. from M: m = <1,-3,-1> & from Q: e = <2,-1,-1> where m ⊥ M  &  e ⊥ Q

   2.  definition: m o e =  √[1+9+1] · √[4+1+1] · cos θ  =  √66  cos θ

   3.  theorem:   m o e =  [1][2] + [-3][-1] + [-1][-1] = 2+3+1  = 6

   4. so  √66 cos θ = 6  thus  θ = cos⁻¹ [6/√66]    Or   θ = cos⁻¹ [√66 / 11]

$$\theta = \cos^{-1}[6/\sqrt{66}] \quad Or \quad \theta = \cos^{-1}[\sqrt{66}/11]$$

II. Find the distance , d, from pt P ( 1, 1, -3) to pl. M: 2x - y + 3z = 6

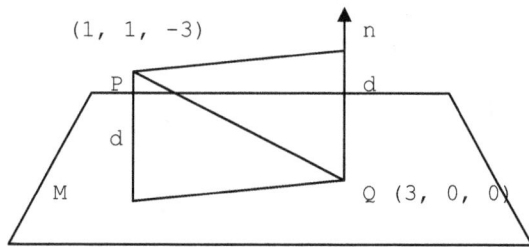

   1. find any point Q in M by letting, say ,y = z = 0 in M; then, x = 3: Q (3,0,0)

   2 . n = <2, -1, 3> ⊥ M   ; take the n whose initial point is Q

   3.  the vector QP = <1-3, 1-0, -3-0> = <-2, 1, -3>

   4. from 5.2.D.: proj ₙ QP = [n o QP / |n|² ] n =

                      = [{-4 - 1 - 9} / {4 + 1 + 9} ]  <2, -1, 3>

                      = <-1, -2, -2>

   5. d = |proj ₙ QP| = |<-1, -2, -2>| = √ [1+4+4]    = 3k

   6. distance d = 3.

6.18 Find the equation of a plane Q that passes through the points

        A(2, -1, -1) & B(1, 3, -1) which is also Parallel to the line of intersection,

   L, of the planes R,  x - 2y + z = 1,   & S, 2x+ 3y -2z = 0.

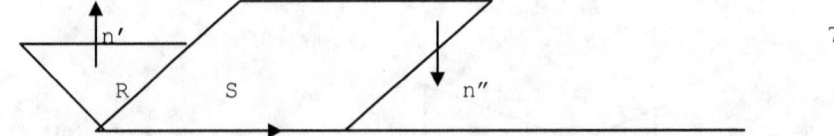

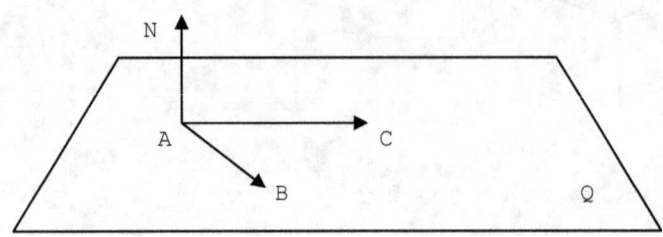

1. n' = <1, -2, 1> ⊥ R    &        n" = <2, 3, -2> ⊥ S

2. so  n' ⊥ L  &  n" ⊥ L

3. thus  L ⊥ the plane determined by  n' & n"  and

   L ⊥ [n' x n"]   ; let n' x n" = C

where n' x n" = <1. -2, 1> x <2, 3, -2> = $\begin{vmatrix} i & j & k \\ 1 & -2 & 1 \\ 2 & 3 & -2 \end{vmatrix}$ = i + 4j + 7k

   so C = <1, 4, 7>

4. take the  C so that its initial point is A; thus C is in plane Q,

5. in plane Q : form AB = <-1, 4, 0>

6. find N = AB x C  = $\begin{vmatrix} i & j & k \\ -1 & 4 & 0 \\ 1 & 4 & 7 \end{vmatrix}$  = <28i + 7j - 8k>  where N ⊥ Q

7. plane Q : 28 [ x - 2 ] + 7[ y + 1 ] - 8[ z + 1 ] = 0

8. plane Q : 28x + 7y - 8z = 57

6.19. Equations of Lines in 3-space:

   Given: $P_1$ ($x_1$ ,$y_1$ ,$z_1$ )  & $P_2$ ($x_2$ ,$y_2$ ,$z_2$ ) two points in 3-space,

   Find:  Equation of the line L  determined by points $P_1$  and  $P_2$   .

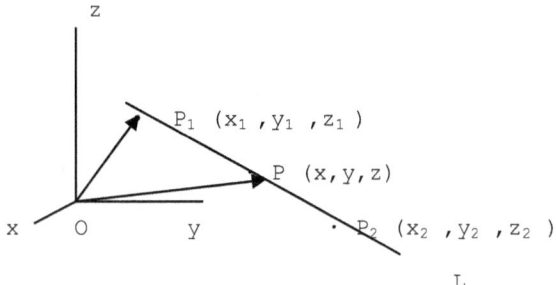

I.                                                    L

1. Let pt P $(x, y, z)$ be any point on $P_1 P_2$

   2. draw OP and label it V

   3. $V = OP_1 + t P_1 P_2 : t \varepsilon \mathbf{R}$      is The Vector Equation of the line $P_1 P_2$.

II.

1. Using point P    I.3. can be written:

   $$\langle x, y, z \rangle = \langle x_1, y_1, z_1 \rangle + t \langle x_2 - x_1, y_2 - y_1, z_2 - z_1 \rangle$$

   $$= \langle x_1 + t(x_2 - x_1), y_1 + t(y_2 - y_1), z_1 + t(z_2 - z_1) \rangle$$

   2. so :  $\left.\begin{cases} x = x_1 + t(x_2 - x_1) \\ y = y_1 + t(y_2 - y_1) \\ z = z_1 + t(z_2 - z_1) \end{cases}\right\}$    the Parametric Equations of L
   with the Parameter t        .

   3. this is not an Unique form because we could have set it up with point $P_2$

   to get:  $\left.\begin{cases} x = x_2 + t(x_2 - x_1) \\ y = y_2 + t(y_2 - y_1) \\ z = z_2 + t(z_2 - z_1) \end{cases}\right\}$    also
   the Parametric Equations of L
   with a Parameter t

III.

   1.  for a simpler form that does not have a different, special name:

   let $x_2 - x_1 = a$; $y_2 - y_1 = b$; $z_2 - z_1 = c$  where $\langle a, b, c \rangle \parallel P_1 P_2$

   2.  so II.2. changes to  the Parametric Equations of $P_2 P_2$ :

   $\left.\begin{cases} x = x_1 + t\,a \\ y = y_1 + t\,b \\ z = z_1 + t\,c \end{cases}\right\}$  Or  $\left.\begin{cases} x = x_2 + t\,a \\ y = y_2 + t\,b \\ z = z_2 + t\,c \end{cases}\right\}$  $\langle a, b, c \rangle \parallel P_1 P_2$  .

IV.

    1. take each equation in III.2. and solve for t:

        $t = [x - x_1]/ a$   &   $t = [y - y_1]/ b$   &   $t = [z - z_1 ]/c$

    2. so     $[x - x_1]/ a$    =    $[y - y_1]/b$    =    $[z - z_1]/c$

          the Symmetric or Standard Form of a Straight Line, L, where

          $(x_1 , y_1 , z_1 )$ is any point on L and $<a, b, c > \| L$

    3. again this is not unique because it depends on what point is used on L.

    4. the  a,b,c are called the Direction Numbers of L.


6.20 Examples:

  I. Find one set of parametric equations of PQ with P (7, -3, 5) & Q (-2, 8, 1).

  Where does this line pierce the xy-plane?

  1. PQ = <-2 - 7, 8 + 3, 1 - 5>   =   <-9, 11, -4>

  2. PQ || PQ; so the direction numbers are -9, 11, -4 ,

  3. so PQ :  x =  7 -  9t

          y = -3 + 11t

          z =  5 -   4t

  4. PQ pierces the xy-plane where z = 0; so  0 = 5 - 4t implies  t =5/4,

    thus   x = 7 - 9 [5/4] = -17/4    &   y = -3 + 11 [5/4] = 43/4

  5. piercing point is ( -17/4, 43/4, 0 ]


II. Find the equation of the line of intersection of the two planes:

               P: 3x + 7y + z = 14     & Q: 4x + y -2z = 2

  1. need two points common to both planes; so solve the system [more methods later]

      3x + 7y +  z  = 14           or         6x + 14y +  2z  = 28

      4x +  y  - 2z  = 2                4x +   y  - 2z   =  2

  2. add:  10x  + 15y + 0 = 30      or    y = 2 - 2x/3

    so  3x  +  7[2 - 2x/3] + z = 14    or    z = 5x/3

3. choose two convenient, arbitrary x's; as:

   let x = 3 , then y = 0  &  z = 5     as point A

   let x =-3 , then y = 4  &  z = -5    as point B

4. AB = <-3 - 3, 4 - 0, -5 -5>  = <-6, 4, -10> = -2 <3, -2, 5>

   can use 3, -2, 5 as a set of direction numbers,

5. One Symmetric form of the line of intersection:

   $$[x - 3]/3 = y/-2 = [z - 5]/5$$

III. Prove the line L, $[x + 3]/-1 = [y - 1]/4 = [z - 2]/1$ , is parallel

   to the plane P,  2x + y - 2z = 4 .

1. now:  n = <2, 1, -2> ⊥ P ; so all vectors ⊥ n are  || to plane P

2. the  vector A = <-1, 4, 1>   is  || L and thus along L,

3. A o n = <-1, 4, 1> o <2, 1, -2>  = -2 + 4 -2 =0  ; so  A ⊥ n

4. therefore  line L || plane P.


6.21 Triple Vector Products.

   I.

       1. they will have this form:  [A x B] x C   &  A x [B x C]

       2. They will not be equal in general.

   II. Prove :  [A x B] x C = [A o C]B  -  [B o C]A  where no vector is Zero.

   Case I. Where  A || B:

       1. then B = s A    s ≠ 0

       2. left side: [A x B} x C = [A x s A] x C = s [A x A] x C = s[0] x C = 0

          right side: [A o C]B - [B o C]A = [A o C]s A - [s A o C]A

                                      = s { [A o C]A - [A o C]A } = 0

       3. so   [A x B] x C = [A o C]B  -  [B o C]A

   Case II. Where  A  is not parallel to B:

       1. Cross Products give Normal Vectors; so in the figure :

          let A x B= n'  and n' x C = n"

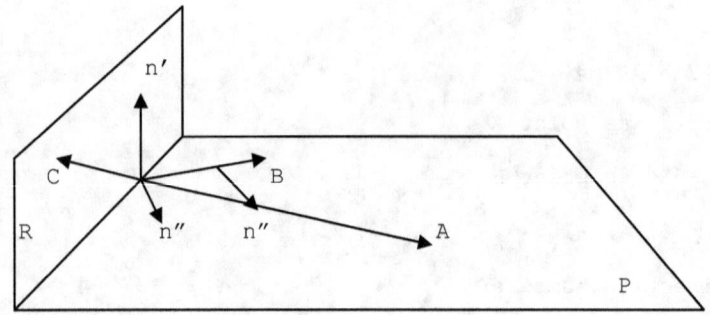

2. and: A & B determine a plane P; so n' ⊥ P

3. n'  & C determine a plane R; so  R ⊥ P  { if a line is ⊥ to a plane then any

plane containing the line is ⊥ to  the first plane.}

4. n" ⊥ R; so n" is in Plane P  { if two planes are ⊥ , then any line ⊥ to one lies

in the other.}

5. now take n" as the vector in p whose initial point is on the B vector and whose

terminal point is on A; so we can write   n" =  r A  + t B : r,t real numbers.

6.  so  [A x B] x C = r A + t B

7. Form a new Basis for A, B, C in order to simplify the calculations:

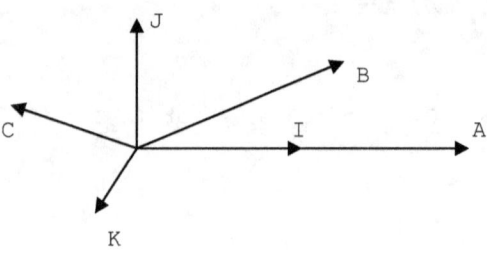

a. let I  = A / |A|    so |I| = 1,

    let J be I rotated by 90°  in the plane of A & B; so |J| = 1, J ⊥ I,

    let K  = I x J ; so |K| = 1 , K ⊥ I  &  K ⊥ J

b. thus A = $a_1$ I                              $a_1$  ≠ 0

        B = $b_1$ I + $b_2$ J                      $b_1$  & $b_2$ ≠ 0

        C = $c_3$ K + $c_1$ I + $c_2$ J              $c_1$ , $c_2$ , $c_3$ ≠ 0

8.
$$[A \times B] \times C = \begin{vmatrix} I & J & K \\ a_1 & 0 & 0 \\ b_1 & b_2 & 0 \end{vmatrix} \times C = a_1 b_2 \ K \times C$$

thus $[A \times B] \times C = \begin{vmatrix} I & J & K \\ 0 & 0 & a_1 b_2 \\ c_1 & c_2 & c_3 \end{vmatrix} = I[- a_1 b_2 c_2] + J[a_1 b_2 c_1] + K[0]$

and $[A \times B] \times C = I[- a_1 b_2 c_2] + J[a_1 b_2 c_1]$

9. into step #6:

$$I[- a_1 b_2 c_2] + J[a_1 b_2 c_1] = r A + t B$$
$$= r a_1 I + t b_1 I + t b_2 J$$
$$= I [r a_1 + t b_1] + J [t b_2]$$

10. or $- a_1 b_2 c_2 = r a_1 + t b_1$ and $a_1 b_2 c_1 = t b_2$

11. so $t = a_1 c_1$                      thus $t = A \circ C$

and $r = - [b_2 c_2 + b_1 c_1]$          thus $r = - B \circ C$

12. into #6:    $[A \times B] \times C = [A \circ C]B - [B \circ C]A$    so proved

13. Note { where the Dot is Commutative the Cross is Anti-Commutative}:

a. changing #11; $[B \times C] \times A = [B \circ A]C - [C \circ A]B$

b. Commutating: $A \times [B \times C] = - \{ [A \circ B]C - [A \circ C]B \}$


14. or: $A \times [B \times C] = [A \circ C]B - [A \circ B]C$  another form of the Triple Product


6.22 Example: Find a way to express $[A \times B] \times [C \times D]$ as a sum:

1. let $C \times D = V$ then $[A \times B] \times [C \times D] = [A \times B] \times V$
$$= [A \circ V] B - [B \circ V] A$$
$$= [A \circ (C \times D)]B - [B \circ (C \times D)]A$$

2. this can be written as: $[A \times B] \times [C \times D] = [A \circ C \times D]B - [B \circ C \times D]A$ , because the ( ) are not necessary to tell us what has to be done first. A Dot gives a scalar which can not be Crossed with a vector; so Cross Products have to be done first.

6.23  Construct the Triple Vector Product [A x B] x [C x D] .

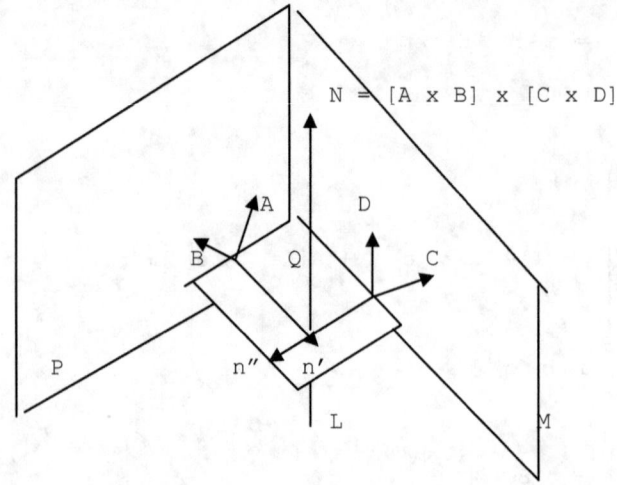

N = [A x B] x [C x D]

1. A & B determine a plane P ; C & D determine a plane M ; P & M intersect in line L

2. let A x B = n'   and  C x D = n"

3. n' & n" determine a plane Q

4. pl.Q ⊥ pl. P   & pl. Q ⊥ pl. M   ; P & M intersect in line L

5. [A x B] x [C x D] = n' x n'' = N as indicated in the figure;

   and N ⊥ Q   and N || L


6.24 Triple Scalar Product

I.

1. In electrical engineering studies as in many other studies the vector notation has proved to be very useful. If we  measure changes in Electromotive Force by dE and induced changes in the distance moved by a point along a conducting wire by dl where v  is the velocity, then dE = [ B x dl ] o v where B is the Flux Density for the magnetic Field the point moves through.

2. Here we are interested in the mathematics of this form and not the physics. The example was used just to point out where the form might have use.

3. The form [A x B] o C is called a Triple Scalar Product because we can see that the final value would be a scalar although it involves products with three vectors.

4. Some writers call the form a Scalar Triple Product - we will not do this.

II.

1. Form a Parallelepipied using A & B as the adjacent sides of the Base whose plane is P and C as the adjacent edge.

2. take A x B = N where N ⊥ P

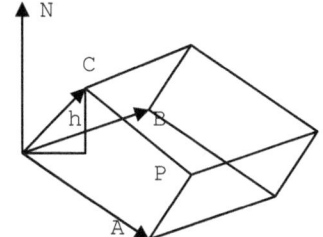

 3. [A x B]  C = |A x B| |C| cos θ  where , 0< θ < π/2 ,is the angle between N & C,

 4. from the terminal point of C drop a ⊥ to P to get h = |C| cos θ ; this h is the Altitude of the Parallelepipied,

 5. |A x B| is the area of the Base from a previous theorem,

 6. from geometry : [A x B] o C = + or - The Volume, V, of the Parallelepipied.

7.

   a. the minus possibility occurs because C might be oppositely directed as shown in this graph.

   b. if A, B, C form a Right Hand System then we take the +; while

       if A, B, C form a Left Hand System then we take the - .

8. from what we said in the last section this product could be written as:

       A x B o C  = + or  - V

III.

1. from geometry the Volume will not change if we consider that B & C form the Base and A the lateral edge, and so on;

2. A x B o C = B x C o A = C x A o B , but we  must keep the Same-Handed System through out the chain of equals .

IV.

1. The Dot product is commutative ; so  $[A \times B] \circ C = [B \times C] \circ A$

   can be changed to:                    $[A \times B] \circ C = A \circ [B \times C]$

2. or  $A \times B \circ C = A \circ B \times C$

3. In words in the triple scalar product we may interchange the Dot and the Cross as long we Dot between two vectors and Cross between two vectors but not Dot or Cross between a Scalar and a Vector.


V. A Theorem.

1. Given: $A = <a_1 , a_2 , a_3 >$, $B = <b_1 , b_2 , b_3 >$, $C = <c_1 , c_2 , c_3 >$

2. $B \times C = \begin{vmatrix} i & j & k \\ b_1 & b_2 & b_3 \\ c_1 & c_2 & c_3 \end{vmatrix} = i[b_2 c_3 - b_3 c_2 ] - j[b_1 c_3 - b_3 c_1 ] - k[b_1 c_2 - b_2 c_1 ]$

3. $A \circ [B \times C]$

   $= [a_1 i + a_2 j + a_3 k] \circ i [b_2 c_3 - b_3 c_2 ] - j [b_1 c_3 - b_3 c_1 ] - k [b_1 c_2 - b_2 c_1 ]$

   $= a_1 b_2 c_3 - a_1 b_3 c_2 - a_2 b_1 c_3 + a_2 b_3 c_1 + a_3 b_1 c_2 - a_3 b_2 c_1$

   $= a_1 b_2 c_3 + a_2 b_3 c_1 + a_3 b_1 c_2 - a_3 b_2 c_1 - a_1 b_3 c_2 - a_2 b_1 c_3$

4. $A \circ B \times C = A \times B \circ C = \begin{vmatrix} a_1 & a_2 & a_3 \\ b_1 & b_2 & b_3 \\ c_1 & c_2 & c_3 \end{vmatrix}$

VI.

   1.  $A \circ B$  is a scalar , say, s

   2.  $[A \circ B] C = s C$ ;  so $[A \circ B] C \parallel C$

VII. another Theorem.

1. $[A \times B] \circ [C \times D] = \{[A \times B] \times C \} \circ D = \{ [A \circ C]B - [B \circ C]A \} \circ D$

2. $[A \times B] \circ [C \times D] = [A \circ C] [B \circ D] - [B \circ C] [A \circ D]$

VIII. Example.

Given: the Position vectors $A = <1,-1,1>$, $B = <2,3,-1>$, $C = <-1,2,2>$

Find: all points P [x,y,z] in the plane Q such that $|OP| = 1$ & OP $\perp$ C

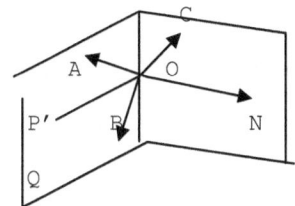

1. Find a vector N ⊥ plane Q

2. N = A x B =
$$\begin{vmatrix} i & j & k \\ 1 & -1 & 1 \\ 2 & 3 & -1 \end{vmatrix}$$
= <-2,3,5>

3. take point P' in Q so that OP' ⊥ N

4. we want OP' to be ⊥ C so that OP' is ⊥ plane of N & C

5. thus OP' = N x C =
$$\begin{vmatrix} i & j & k \\ -2 & 3 & 5 \\ -1 & 2 & 2 \end{vmatrix}$$
= <-4,-1,-1>

6. the unit vector OP = OP'/|OP'| =    <-4,-1,-1>/ √[16+1+1]

   or                      = < - 4/3√2 , - 1/3√2 , - 1/3√2 >

7. thus x = - 4/3√2  ,  y = - 1/3√2  , z = - 1/3√2

8. point $P_1$ = (- 4/3√2, - 1/3√2, - 1/3√2)

   It is clear that point $P_2$ = (4/3√2, 1/3√2, 1/3√2) also satisfies the conditions of

the problem.


6.25.  Skew Lines.

   I.

   1. Definition: Skew lines are a pair of different lines that are not in the same

plane.

   2. Two important properties of skew lines:

      a. they can not determine a plane,

      b. they do not intersect.

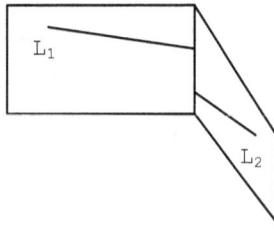

II.

1. We have been using vector methods to develop some theorems about geometric figures. In some cases a student may feel that we have taken too many liberties with rigor in dealing with vectors as representing [geometric] lines, and of course we have not.

2. Consider the figure:

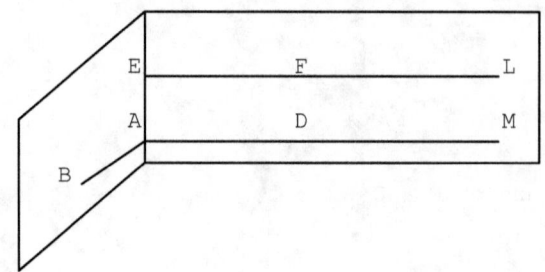

3.  Lines L  & M  are different geometric figures, but vectors EF & AD are mathematically the same vectors even when representing the two lines.

4. Point A can be considered to be on both vectors but is not on both lines.

5. Where non-intersection of lines is important, the study of skew lines graphically points out this "flaw".

6. EF || AD; so they determine a plane. While they do not intersect, they are in the same plane. EF & AD are not skew lines.

7. AB & AD are not in the same plane, but they do intersect; so they determine a plane that is not shown.  AB & AD are not skew lines.

8. AB & EF are not in the same plane, and they do not intersect. AB & EF are skew lines.

III.

1. If |AD| = |EF|, then as vectors AD = EF; so we really can not consider that AB & EF are skew vectors for they share the common point A. In the Geometry of Lines the two do not have a common point A, but in the Geometry of Vectors they do.

2. If this  situation was handled by purely algebraic means, this  dual mathematics would probably not be obvious.

3. Parallelism in Vector theory is an Equivalent/Equal Class , but in Geometric Line Theory it is not.

4. The reason this duality is slurred over in most writings is because as long as we stick to Vector Theory we will have no inconsistencies.

5. So, while the duality can be ignored in most cases, this  discussion should answer some of the unasked questions that sometimes arises in the minds of some students.


6.26. Definition of the Distance between two Skew Lines.

I.

1. Consider the bundle of planes [Q, R, S, P, ... ] whose line of intersection is n which is one of two Skew lines, and plane Q as the one that is perpendicular to the other Skew line, m  ,

2. the figure shows how this is possible. We are looking straight down the line n; so it appears to be just a point. Each plane appears to be a line through that point.

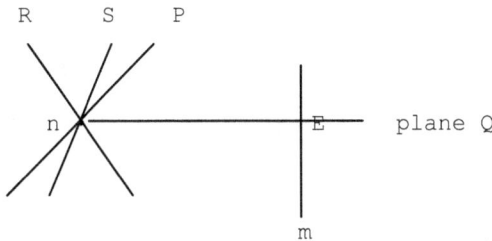

3. take E as the point of intersection of the plane Q and line m  ,

4. in pl.Q drop a perpendicular, p , to n  from pt. E; this intersects n at  pt. F ; as:

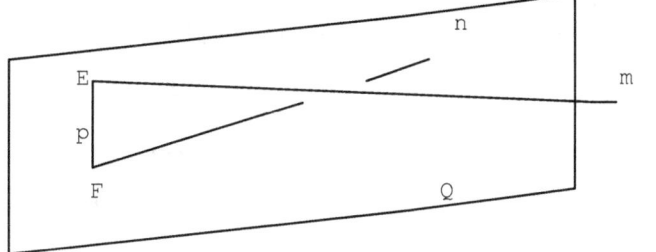                                        so so p ⊥ n at point F,

5. and  p ⊥ m  at pt. E  [ if a line is ⊥ to a plane, it is ⊥ to all lines in the

plane through the foot of the ⊥ ] ,

6. we Define the length of the line segment,FE, as the Distance ,d, between the two

Skew Lines; this does seem to represent the shortest distance between the two lines.

II.

   1. It seems that in general in 3-space it would be difficult to decide whether two

given lines are skew lines or whether in fact they intersect. Off-hand we do not

seem to have either a geometric nor an algebraic method to determine skewness.

However, this distance work gives us a method.

   2. Follow the steps of the definition and if d =0 then the lines are not skew, and

they do intersect at that point where the distance d = 0.

   3. If A & B represent two lines L  & M  and A = s B, the the two lines are not

skew for they are parallel.

   4. If  A  ≠  s B  and d  ≠  0 , then L  &  M  are skew.

III. Exercise: Find the distance between AB where A = (-2,3,-4) and

B =(1,2,7) & CD where C = (5,7,-3) and D = (2,-1,4).

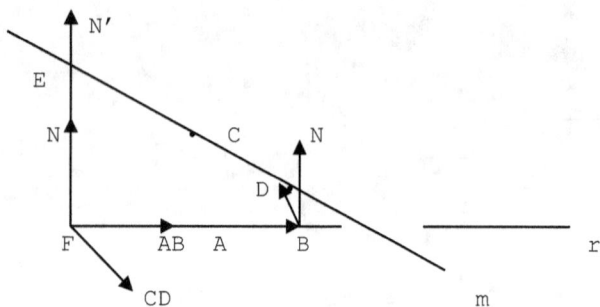

1. CD = <-3,-8,7>   & AB = <3,-1,11>

2. take the equivalent CD & AB that have the same initial point F. As far as the

algebra is concerned previous work indicated that this visualization is not

necessary , but it does help us fill in the gaps in the geometry.

$$3. \text{ let N' = CD x AB} = \begin{vmatrix} i & j & k \\ -3 & -8 & 7 \\ 3 & -1 & 11 \end{vmatrix} = <-81, \ 54, \ 27> = 27 <-3,2,1>$$

4. Let N = N'/27 to make the calculations easier; so N = <-3,2,1>  where

  $N \perp CD$   &   $N \perp AB$

5. from previous and the figure : d =  Proj $_N$  BD  [ note AC or AD or BC could have

b      been used ]

6. BD = <1, -3, -3> so

  d =  [BD o N] / |N|  = <1,-3,-3> o <-3,2,1> / | N | = [-3-6-3]/ √ [9+4+1] =

  d  = -12/√14


IV. Find the distance between the lines AB & CD where the points are

  A (1,0,1) , B (2,0,1) , C(0,1,1) , D(0,2,1).

1. AB  = <1,0,0>  &  CD = <0,1,0>

$$2.\ N = AB \times CD = \begin{vmatrix} i & j & k \\ 1 & 0 & 0 \\ 0 & 1 & 0 \end{vmatrix} = <0, 0,1>$$

3. choose as the cross vector BD = <-2, 2, 0>

4. d = [ BD o N ] / | N | = [<-2,2,0> o <0,0,1>] /|N| = [0+0+0] / |N| = 0

5. d = 0    so   not Skew; so there is no distance between the lines.


6.27 The Field of Complex Numbers , Z .

  I. This is a final say about sets of vectors used as the Domain of a  function. The

  statements are for information rather than for use. The Real numbers will still be

  the basis for all of our number situations in this text.

  II.

   1. A Complex Number, z , is defined as: z = x + i y where x,y are Real Numbers and

  i, the Imaginary Unit, is by defined by  $i^2$ = -1 ; it is  symbolized by  i = √-1.

   2. Its geometry is:

  [the imaginary Axis]

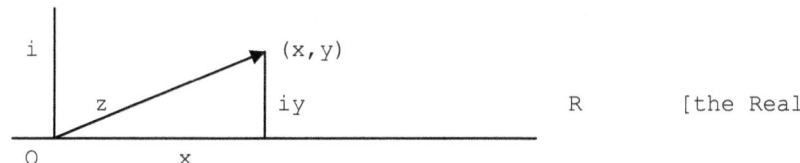

R      [the Real Axis]

geometrically the complex number z is a vector whose equivalent/equal class is the ordered pair of numbers (x, y).

3. Given $z_1 = x_1 + i y_1$ & $z_2 = x_2 + i y_2$ ,

then $z_1 = z_2$ iff $x_1 = x_2$ & $y_1 = y_2$

III. From previous material it is clear that the following are true:

1. $z_1 + z_2$ is an unique number in Z

2. $z_1 + (z_2 + z_3) = (z_1 + z_2) + z_3$

3. $z_1 + z_2 = z_2 + z_1$

4. O is the Additive Identity Number for $z + 0 = 0 + z = z$

5. $-z$ is the Additive Inverse [or Negative] for

$-z \equiv -x - i y = 0$

6. $k z \equiv k x + i k y$ is a unique number in Z

7. $k [z_1 + z_2] = k z_1 + k z_2$

8. $[k + m] z = k z + m z$

9. $k [m z] = [k m] z = k m z$

10. $1 z = z$

IV. Z is an Abstract Vector Space; so we can use any complex number z in the same manner that we used vectors in the previous material. Many theorems can be proven about complex numbers, and they have great use. There is not time nor space to go into that material in this text.

V. We will do this however --- prove the function $f(z) = 5 z$ is linear:

1. $f(z_1 + z_2) = 5 [z_1 + z_2) = 5 z_1 + 5 z_2 = f(z_1) + f(z_2)$,

2. therefore this function of a complex number is Linear.

CHAPTER SEVEN. MATRIX THEORY

7.0 Introduction.

In order for a function to be linear we have to know something about addition of its independent variables. The way that we handle Linearity we would also need to know something about the scalar product of its independent variables. An Abstract Vector Space deals with these two concepts; so a likely choice for a Linear Function is one whose domain is a Vector Space.

Looking over some of the work that has been done recently involving Matrices one might guess that matrices form a Vector Space although the theorems were probably not handled in that manner. The form y = A x does come up in some of this work , and we have proved that that form is linear --- even though the y, A, & x have different meaning in matrix theory  than they did in our proof. In matrix theory some work in products gives a linear form. Thus, it seems reasonable to study Matrix theory in terms of linearity at this time.

Certain of our previous problems involved solving systems of simultaneous first degree equations. To solve these systems we fell back on some of the simple devices used in elementary algebra courses. A more serious study of general methods of solutions lead in the middle 1800's to an object called a Matrix. We will now enlarge our list of domains of Linear functions by studying the mathematics of matrices.

The concept was born in the middle 1800's but not put to general use until 1929.

7.1 Matrices in general.

1. Matrix theory is an elegant and powerful mathematical language that can be used in the study of systems of equations and in the study of  physics, engineering, statistics, and so on. In such studies a certain pattern involving numbers occurred enough times to warrant special treatment. A nomenclature and symbolic notation was adopted and using basic definitions a Mathematics was developed by inventing theorems and working out their proofs.

However, to this day the nomenclature and the symbols have not been standardized, and this complicates somewhat the study of matrices in the minds of  beginning students. In this text we will use what seems to be the most common nomenclature and simplify the symbols as much as possible.

2. This pattern was called a Matrix , and its development into a mathematics

simplified previous studies by eliminating the superfluous. This work also unified

considerably our previous treatment of number expressions that had this pattern.

   3. In Analytic Geometry the coordinates of a point P(x,y) are changed under

rotation of axes to P'(x',y') by a system of equations of the form   x' = A x + B y

                                                                      y' = C x + D y

We note that this system is completely determined by the pattern $\begin{bmatrix} A & B \\ C & D \end{bmatrix}$  ;

so working with just this block of numbers could simplify any work we do with

the equations in terms of the new axes.

   4. Now a matrix is a way of displaying information about its numbers. It does not

represent a number; thus it has no number value itself. We will learn to use a matrix

as if it were a number. We will manipulate this display of numbers in an algebraic

manner that allows us to find consequences of the original information in a much easier

manner than by using older methods.


7.2 Definition of a Matrix noting that it has Rows and Columns as its crucial parts.

   1. A   m x n Matrix  is a rectangular array of numbers, real or complex, arranged

in m rows and n columns surrounded by a set of brackets [some writers use parenthesis].

       Example: $\begin{bmatrix} 2 & 0 & -1 \\ 3 & \frac{1}{2} & 3 \end{bmatrix}$     is a matrix.

   2. The following notations are used to symbolize a matrix; this can also be thought

of as a definition:

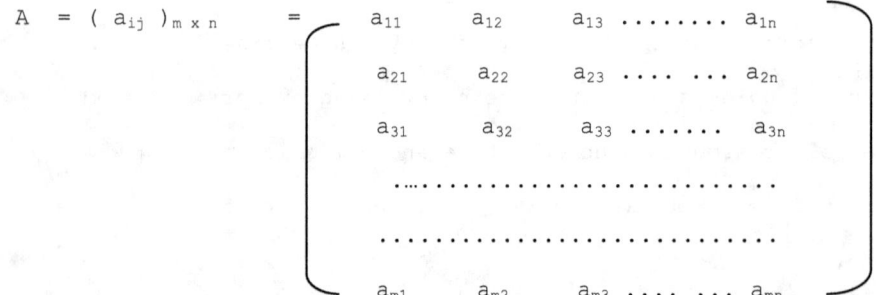

the    $a_{ij}$   are the Elements   or  Entries of the matrix and are written here in a

double subscript form where the first subscript is the number of the row and the

second is the number of the column. This is for the convenience of locating the

Position or Address of a particular entry which is very important in most but not of our work; so we could use a, b, c, ....,x as entries in some of our work. In all of our work with matrices we list the row first. It is clear that  i = 1, 2, 3, ... m & j = 1, 2, 3,  ... n. It is best to get the 1st row first by fixing i on 1 and letting j vary 1 through n; then fixing  i on 2 as j varies 1 to n for the second row; and so on to j = n.

3. The elements $a_{ij}$   are scalars in this text. The Dimension or Order of this matrix is m x n and is read   "m by n" --- it has a   m row order and a    n column order.

4. Some writers use the A in heavy type or in script form; we will not do that because those forms can not be handwritten.

7.3

1. The special matrix  ( a  b  c  ... x ) is called a Row Matrix, and this would be obvious in the double subscript form : ( $a_{i1}$    $a_{i2}$    $a_{i3}$ ... $a_{in}$ ).  Note: the i th row.

2. There are short forms that are used; as: ( $a_{ij}$ )$_{m \times n}$  where m = 1, or ( $a_{ij}$ )$_{1 \times n}$ ,  or ( $a_{ij}$ )$_{4 \times n}$ ,  or  ( $a_i$ )$_n$        , for a row matrix.

3. This reminds us that vectors can be written as  A = <$a_1$ ,$a_2$ ,$a_3$ ,...,$a_n$ >  which is in the form of a row matrix. The commas in the vector form and their lack in the matrix form have no mathematical significance; they are in the vector form because vectors are equivalent/equal to points ( x, y). So, a Row Matrix is a Vector and a Vector is a Row Matrix. Without further proof the properties [ where they make sense ] of vectors are the properties of row matrices. When we find properties of row matrices, we see that they will also apply to vectors [ where they make sense ]. The distinction between A for a matrix and A for a vector no longer needs to be made; we can use A for both.

4. The matrix      $\begin{pmatrix} b_{1j} \\ b_{2j} \\ b_{3j} \\ . \\ b_{mj} \end{pmatrix}$   is called a Column Matrix  or a Column Vector.

some write this horizontally to save space; as: ( $b_{1j}$  $b_{2j}$  $b_{3j}$  ... $b_{1j}$ )         where it is clear from the subscripts that it is a column matrix.

5. A whole matrix  A  = ( $a_{ij}$ )$_{m \times n}$    can be thought of as an Ordered Set of Row Vectors or as an Ordered Set of Column Vectors.

## 7.4 A Square Matrix:

1. The matrix $A = (a_{ij})_{n \times n}$ is a Square Matrix where n is its order.

2. The set of elements $\{ a_{11},\ a_{22},\ a_{33},\ldots\ldots\ldots, a_{nn} \}$ in matrix A is called the Principal or Primary or Main Diagonal and the reverse set $\{ a_{1n},\ a_{2\ n-1},\ a_{3\ n-2},\ldots\ldots\ldots, a_{n\ 1} \}$ is the Secondary Diagonal. Note: a Set is a collection of numbers in any order separated by commas and enclosed by { }; note this is not [ ] nor ( ) which is matrices' method of bracketing. In general we prefer to list the items in a logical/numerical order.

## 7.5 The Basic Concepts of Matrices.

### I. Submatrix of a Matrix.

1. If we delete row(s) and/or column(s) without any other changes in a matrix, the resulting matrix is a Submatrix. Note: the symbol row(s) means either one row or many rows.

2. For convenience we consider that a matrix is a submatrix of itself.

3. Illustration: all possible submatrices of $\begin{bmatrix} a & b \\ c & d \end{bmatrix}$ are:

$$\begin{bmatrix} a & b \\ c & d \end{bmatrix} ; \begin{bmatrix} a \\ c \end{bmatrix} ; \begin{bmatrix} b \\ d \end{bmatrix} ; (a\ b) ; (c\ d) ; (a) ; (b) ; (c) ; (d) .$$

### II. Equality.

1. Despite the fact that matrices are not numbers but rather ways of displaying patterns of numbers or information, we can define the equality between matrices. Two matrices are equal if they display the same information. Basically we are using the Equivalent/Equal Class concept in this definition and in the others that will follow.

2. Given:  $A = (a_{ij})_{m \times n}$  $B = (b_{ij})_{m \times n}$

Then by definition:  $A = B$  iff  $a_{ij} = b_{ij}$  for all  $i = 1,2,3,\ldots,m$

for all  $j = 1,2,3,\ldots,n.$

3. In words two matrices are equal iff :

    a.  they have the same number of rows and the same number of columns,

    b.  all correspondingly placed elements are equal.

4. It is easy to prove that the definition satisfies the four properties of the Equality concept. It is hard to write down the proof because each step is so obvious. Each element is a real number ; so step #3. a,b above  is enough to satisfy each of the four properties.

III. Negative: Given the matrix $A = (a_{ij})_{m \times n}$    , then the Negative of A

is   $-A$   where   $-A = (-a_{ij})_{m \times n}$             ; or

if    $A =$
$$
\begin{pmatrix}
a_{11} & a_{12} & a_{13} \cdots\cdots a_{1n} \\
a_{21} & a_{22} & a_{23} \cdots\cdots a_{2n} \\
\cdots\cdots\cdots\cdots\cdots\cdots\cdots \\
\cdots\cdots\cdots\cdots\cdots\cdots\cdots \\
a_{m1} & a_{m2} & a_{m3} \cdots\cdots a_{mn}
\end{pmatrix}
$$

then   $-A =$
$$
\begin{pmatrix}
-a_{11} & -a_{12} & -a_{13} \cdots\cdots -a_{1n} \\
-a_{21} & -a_{22} & -a_{23} \cdots\cdots -a_{2n} \\
\cdots\cdots\cdots\cdots\cdots\cdots\cdots\cdots \\
\cdots\cdots\cdots\cdots\cdots\cdots\cdots\cdots \\
-a_{m1} & -a_{m2} & -a_{m3} \cdots\cdots -a_{mn}
\end{pmatrix}
$$

IV. The Zero Matrix, 0:

   1. 0 is a matrix whose elements are all 0's ,

   2. it is also called the  Null Matrix ,

   3. so: $0 = (0_{ij})_{m \times n} =$
$$
\begin{pmatrix}
0 & 0 & \cdots & 0 \\
0 & 0 & \cdots & 0 \\
\cdots\cdots\cdots \\
0 & 0 & \cdots & 0
\end{pmatrix}
$$

V. The Unit Matrix,  $I_n$  :

   1. the unit matrix $I_n$   is a Square Matrix whose Main Diagonal is all   1's and all the other elements are  0's.

   2. so   $I_n = (a_{ij})_{n \times n}$    where $a_{ij}$   = 1 for i = j  and  a = 0  for  i ≠ j

$i = 1,2,3, \ldots,n$  &  $j = 1,2,3,\ldots,n$

VI. Some writers introduce a matrix J where every element is 1.


7.6  Operations on matrices --- in order to prove that matrices form a Vector Space we need the following. Each has its own important uses also.

I.

1. Definition: Given  $A = (a_{ij})_{m \times n}$        $B = (b_{ij})_{m \times n}$

note both have the same Row & Column dimensions,

Then  $A + B = (a_{ij})_{m \times n} + (b_{ij})_{m \times n}$

$= ([a + b]_{ij})_{m \times n}$

2. The discussion on the previous use of the word "add" and the symbol " + " in our work in Vectors applies here.

3. Example: if $A = \begin{bmatrix} 1 & 0 & 3 \\ 2 & -1 & 4 \end{bmatrix}$ & $B = \begin{bmatrix} -1 & 0 & 0 \\ 0 & 2 & 3 \end{bmatrix}$

then  $A + B = \begin{bmatrix} 1-1 & 0+0 & 3+0 \\ 2+0 & -1+2 & 4+3 \end{bmatrix} = \begin{bmatrix} 0 & 0 & 3 \\ 2 & 1 & 7 \end{bmatrix}$

II. Some Theorems [ where all matrices are of the same dimension].

These are easy to prove using the properties of real numbers. We will show two such proofs; note while A &  $(a_{ij})_{m \times n}$  are matrices the  "a"  in $(a_{ij})_{m \times n}$    is a Real Number.

1. A + B  Exists and is a Matrix                                    Closure Law

2. A + B = B + A                                                    Commutative Law

Proof: $A + B = ([a + b]_{ij})_{m \times n}) = ([b + a]_{ij})_{m \times n})$

$= B + A$

3. $[A + B] + C = A + [B + C)$

$= A + B + C$                                      Associative Law

4. There is a matrix O such that  $A + O = O + A = A$        the Additive Identity

5. For every matrix, A, there is a Negative, -A, such that

$A + [-A] = [-A] + A = O$                                 the Additive Inverse

Proof: $A + [-A] = (a_{ij})_{m \times n} + (-a_{ij})_{m \times n} = ([a_{ij} + \{-a\}_{ij}]_{m \times n})$

$= ([a - a]_{ij})_{m \times n} = (0_{ij})_{m \times n} = O$

$[-A] + A = (-a_{ij})_{m \times n} + (a_{ij})_{m \times n} = ([\{-a\}_{ij} + a_{ij}]_{m \times n})$

$$= ([-a + a]_{ij})_{m \times n} = (0_{ij})_{m \times n} = 0$$

III. So, the Set of all m x n matrices forms a Commutative or Abelian Group under Addition.

7.7 For convenience in the rest of Chapter 7  All Matrices will be of the Same Dimension; thus $A = (a_{ij})_{m \times n}$ , $B = (b_{ij})_{m \times n}$ , $C = (c_{ij})_{m \times n}$  , and so on.

7.8 Subtraction of Matrices.

    1.

       a. The Difference between two matrices A & B is defined as the Addition of the Negative of the Second Matrix. We will use  " A – B " and state this as A subtract B.

       b. so   A – B = A + [ –B ]

       c. Subtraction is not Commutative.

    2. The following is a set of theorems that are easy to prove . We will prove two.

       I.  – 0 = 0

       II.  A – A  =  0

         proof: A – A = A + [ –A ] = 0    from 7.6  #5 above.

       III. –(–A) = A

         proof:  –(–A) = – $(a_{ij})_{m \times n}$     = $(-[-a]_{ij})_{m \times n}$   = $(a_{ij})_{m \times n}$   = A

      IV. –[A + B] = [–A] + [–B] = –A –B

       V.  if  X + A = B  , then  X = B – A

    3. We note that we have been careful in setting up our definitions so that the theorems are the same as if the matrices A, B, C, .... were real numbers.

7.9  Multiplication or The Product of a Matrix by a Scalar.

   I.

      1. Definition. Given: a matrix $A = (a_{ij})_{m \times n}$.    &  a scalar q,

              Then:  q · A  or  [q] [A] or  qA

$$= ([q\,a]_{ij})_{m \times n}.$$

2. The labels  Scalar Multiplication or Scalar Product  refer to the Vector

concept and only a row matrix or a column matrix  can be called a vector at this time.

   II. Multiplication Theorems --- these again can be easily proved as developed before.

      1. $-A = [-1]\ A$

      2. $q\ A = A\ q$                          Commutative Law

      3. $q[\ A + B\ ] = q\ A + q\ B$                a Distributive Law

      4. $[\ c + k\ ]\ A = c\ A + k\ A$              another Distributive Law

      5. $c\ [\ k\ A\ ] = [\ c\ A\ ]\ k = [ck]A$

                  $= c\ k\ A$                     Associative Law

      6. $1\ A = A$                              1  is the Identity Element

      7. $O \cdot A = O$

   III. Example:  if  $A = \begin{bmatrix} 1 & 0 \\ -3 & 4 \end{bmatrix}$ , then  $\tfrac{1}{2} A = \begin{bmatrix} \tfrac{1}{2} & 0 \\ -3/2 & 2 \end{bmatrix}$


7.10  Matrices as Vectors Spaces.

   1. It is clear that the Set of all m x n Matrices is an [Abstract] Vector Space.

   2. We say that this Vector Space has the Dimension  mxn .

   3. Thus, every Matrix satisfies all the conditions of being a Vector.

   4. So, we can use the Nomenclature and the Symbolic Notation of Matrices and Vectors

interchangeable. When it helps to keep them separate in our minds , we will write a Row

Matrix as ( $a_{i1}$  $a_{i2}$  $a_{i3}$ ... $a_{in}$ )  and a Vector as  $< a_1 , a_2 , a_3 , ...., a_n >$


7.11 Standard Basis.

   I.

      1. A Linear Combination of Matrices is of the form

         $A = q_1\ A_1\ +\ q_2\ A_2 + q_3\ A_3 + .... + q_n\ A_n.$

      2. We see from 4.1. C. 4. that we can now use that proof to prove the above is

Linear where we consider that the A is a function of  $A_1 , A_2 , A_3 , ...., A_n.$

II.

   1. If  A = 0 for all the q's = 0 then the $A_k$  are Linear Indepentent. If at least one q ≠ 0 , then the $A_k$ are Linear Dependent.

   2. The minimum number of linear independent matrices that can be used to form all of the matrices in a vector space by means of linear combinations form a set of matrices that is defined as a Basis for that space. A Standard Basis is the simplest set that can be used to form a Basis. Simplest is determined in terms of what we are going to do with that basis.

   3. We say that the set of matrices that form a basis Span the space.

   4. We note that our definitions are the same as for vectors.


7.12 Illustration.

   1. Consider the set of all square matrices of order 2,

   2. this set is a vector space of dimension 2 x 2 = 4 ,

   3. consider the special set of such matrices [ note each is a subspace ]:

$$E_{11} = \begin{bmatrix} 1 & 0 \\ 0 & 0 \end{bmatrix}, \; E_{12} = \begin{bmatrix} 0 & 1 \\ 0 & 0 \end{bmatrix}, \; E_{21} = \begin{bmatrix} 0 & 0 \\ 1 & 0 \end{bmatrix}, \; \& \; E_{22} = \begin{bmatrix} 0 & 0 \\ 0 & 1 \end{bmatrix}$$

note: in $E_{bd}$   the bth row & the dth column is a 1 and all other elements are 0's.

   4. their Linear Combination will be :

$k_1 \; E_{11} \; + k_2 \; E_{12} \; + k_3 \; E_{21} \; k_4 \; E_{22} \; =$

$$= k_1 \begin{bmatrix} 1 & 0 \\ 0 & 0 \end{bmatrix} + k_2 \begin{bmatrix} 0 & 1 \\ 0 & 0 \end{bmatrix} + k_3 \begin{bmatrix} 0 & 0 \\ 1 & 0 \end{bmatrix} + k_4 \begin{bmatrix} 0 & 0 \\ 0 & 1 \end{bmatrix}$$

$$= \begin{bmatrix} k_1 & 0 \\ 0 & 0 \end{bmatrix} + \begin{bmatrix} 0 & k_2 \\ 0 & 0 \end{bmatrix} + \begin{bmatrix} 0 & 0 \\ k_3 & 0 \end{bmatrix} + \begin{bmatrix} 0 & 0 \\ 0 & k_4 \end{bmatrix} = \begin{bmatrix} k_1 & k_2 \\ k_3 & k_4 \end{bmatrix}$$

5. as the k's take on all possible values from the real numbers, this linear combination clearly forms all possible 2 x 2 matrices.

6. we also see that $k_1 \; E_{11} \; + k_2 \; E_{12} \; + k_3 \; E_{21} \; k_4 \; E_{22} \; = \begin{bmatrix} k_1 & k_2 \\ k_3 & k_4 \end{bmatrix} = 0$

iff $k_1 \; = k_2 \; = k_3 \; = k_4 = 0$ ; so the E's are Linear Independent,

7. The Set of E's is a Basis. They are so simple they seem to form a Standard Basis.


7.13  A special matrix --- the Transpose.

I. Definition.  Given:  Matrix $A = ( \; a_{ij} \; )_{m \times n}$.

Then:  The Transpose of A, $A^T$ , is defined by

$$A^T \; = \; ( \; a_{ji})_{n \times m} \; .$$

1. some writers use A' for A   because of the ease of typing; we will not do that here.

2. so, $A^T$ is A with the Rows and the Columns interchanged.

II. It is clear that :  $[A^T]^{\;T} \; = A$

II I. Theorem.  Given: $A = ( \; a_{ij} \; )_{m \times n} \quad \& \quad B = ( \; b_{ij} \; )_{m \times n}$.

Then:  $[A + B]^T \quad = A^T \; + B^T$

Proof:

1. $[A + B]^{\;T} \; = [( \; a_{ij} \; )_{m \times n} + ( \; b_{ij} \; )_{m \times n} ]^{\;T} \; = [\{(a + b)_{ij}\}_{m \times n} \; ]^{\;T}$

$= [(a + b)_{ji} \; )_{n \times m}$

$= ( \; a_{ji} \; )_{n \times m} \; + \; ( \; b_{ji})_{n \times m}$

2. $[A + B]^{\;T} \; = A^T + B^T$

IV.Theorem.  Given:  $A \; = ( \; a_{ij} \; )_{m \times n}$

Then: $[ \; q \; A \; ]^{\;T} \quad = q \; A^T$

Proof:

1. $q \; A \; = \; [(q \; a_{ij} \; )_{m \times n}$

2. $[ \; q \; A \; ]^{\;T} \; = [(q \; a_{ji} \; )_{n \times m} \quad =q \; [a_{ji} \; )_{n \times m} \; ]$

3. $[ \; q \; A \; ]^{\;T} \; = q \; A^T$

V. Given any matrix A there is one and only one $A^T$ ; so $A^T$  is a function of A as $A^T \; = f(A)$. We have proved that this function of A , $A^T$ , is Linear.

VI. Some examples    given $A = \begin{bmatrix} 1 & 2 & 3 \\ 0 & 1 & -1 \end{bmatrix}$ & $B = \begin{bmatrix} 5 & -1 & -1 \\ -1 & 0 & 2 \end{bmatrix}$ & $C = \begin{bmatrix} 1 \\ 0 \\ -3 \end{bmatrix}$

1. $A^T = \begin{bmatrix} 1 & 0 \\ 2 & 1 \\ 3 & -1 \end{bmatrix}$ ; $B^T = \begin{bmatrix} 5 & -1 \\ -1 & 0 \\ -1 & 2 \end{bmatrix}$ ; $C^T = \begin{bmatrix} 1 & 0 & -3 \end{bmatrix}$

2. $A^T + B^T = \begin{bmatrix} 6 & -1 \\ 1 & 1 \\ 2 & 1 \end{bmatrix}$ ; $A + B = \begin{bmatrix} 6 & -1 & 2 \\ -1 & 1 & 1 \end{bmatrix}$

   note: $[A + B]^T = \begin{bmatrix} 6 & -1 \\ 1 & 1 \\ 2 & 1 \end{bmatrix}$  so we see that

   $[A + B]^T$ does equal $A^T + B^T$

3. $5A = \begin{bmatrix} 5 & 10 & 15 \\ 0 & 5 & -5 \end{bmatrix}$ ; $[5A]^T = \begin{bmatrix} 5 & 0 \\ 10 & 5 \\ 15 & -5 \end{bmatrix}$ ;

   $5[A^T] = \begin{bmatrix} 5 & 0 \\ 10 & 5 \\ 15 & -5 \end{bmatrix}$ ; we see that $[q A]^T$ does equal $q A^T$

7.14 Some Special SQUARE MATRICES; Take $A = ( a_{ii} )_{n \times n}$ .

   I. SYMMETRIC.

      1. A is Symmetric iff $A^T = A$

      2. so $a_{ik} = a_{ki}$ ; thus the matrix is symmetric across the Principal Diagonal.

      3. as in : if $A = \begin{bmatrix} -2 & 4 \\ 4 & 3 \end{bmatrix}$ , then $A_T = \begin{bmatrix} -2 & 4 \\ 4 & 3 \end{bmatrix}$ ; thus  Symmetric.

   II. SKEW-SYMMETRIC.

      1. A is Skew Symmetric  iff  $A^T = - A$

      2. if $a_{ii} = - a_{ii}$ , then $a_{ii} + a_{ii} = 0$  or  $2 a_{ii} = 0$  or  $a_{ii} = 0$

      3. we see that the elements of the main diagonal are all 0  and the symmetric

elements are Negative.

      4. as in : if $A = \begin{bmatrix} 0 & -5 \\ 5 & 0 \end{bmatrix}$ , then $A^T = \begin{bmatrix} 0 & 5 \\ -5 & 0 \end{bmatrix} = - A$ ;  thus Skew-Symmetric

   III. TRIANGULAR.

      1. A  is Triangular iff all the elements Below  [Or  Above] the Main Diagonal are 0.

2. as in : $\begin{bmatrix} 4 & 6 & 1 \\ 0 & -1 & -2 \\ 0 & 0 & 3 \end{bmatrix}$      this is called     UPPER TRIANGULAR

3. and as in : $\begin{bmatrix} 2 & 0 & 0 \\ -1 & 1 & 0 \\ 6 & 3 & 5 \end{bmatrix}$  this is called     LOWER TRIANGULAR

4.   a.  in the   Upper Triangular   $a_{ik} = 0$   for all   $i > k$

   b.  in the  Lower Triangular   $a_{ik} = 0$   for all   $i < k$


## IV. DIAGONAL.

1.   A  is Diagonal  iff  all elements Above and Below the Main Diagonal are 0.

2.   so all   $a_{ik} = 0$ where   $i \neq k$ ;  thus it is both  Upper & Lower Triangular.

3.   as in   $\begin{bmatrix} 1 & 0 & 0 \\ 0 & 2 & 0 \\ 0 & 0 & -5 \end{bmatrix}$

## V. SCALAR.

1. A is  Scalar iff  A  is  Diagonal and all the Diagonal Elements are Equal to each other.

2. so  $a_{ik} = 0$   where   $i \neq k$ and  $a_{ik}$ = the same constant where   $i = k$.

3. the ZERO MATRIX , 0, is a special Scalar matrix.

4. the UNIT MATRIX , $I_n$ , is a special Scalar matrix.


7.15 Expanded use of the Sigma Notation Symbol $\sum$ to represent Summations.

   I. Introduction.

   1. We will be using the following notation ,      $c_{jk} = \sum_{i=1}^{n} ( a_{ji}\ b_{ik} )_{m \times p}$ ,

to describe each element, $c_{jk}$, of a matrix ,$C_{m \times p}$, with m rows & p columns.  "a" is an element of the matrix A = $(a_{ji})_{m \times n}$   and "b" is an element of the matrix  B = $(b_{ik})_{n \times p}$ ,

   2. This Sigma symbol is more complicated than the one used in the Calculus; so it needs to be explained in some detail.

   II. Meanings.

   1. The symbol "$c_{jk}$"  represents the number c that is in the jth row and kth column of C  ,

2.    $\sum$  represents the sum of two or more terms involving an  "a · b" form in each term,

3.    the indices  i, j, k represent a range of integers 1, 2, 3, ...,

4.    "i" is a part of $\sum$ ; the largest number in i's range describes the number of terms in each sum that represents each element of C; this same number describes the number of columns in A and the number of rows in B;  and "i"  must be the middle subscript of the product of a & b ; as: ( $a_{ji}$ $b_{ik}$ )   .

5. the 1st subscript of the "a" [the j] indicates  Rows , and the 1st subscript

   [the m] in the j,k indicator, ( $a_{ji}$    $b_{ik}$ ) $_{m \times p}$ , gives the range of the Rows in C,

6. the 2nd subscript of the "b" [the k] indicates  Columns , and the 2nd subscript

   [the p] in the j,k indicator, ( $a_{ji}$    $b_{ik}$ ) $_{m \times p}$ , gives the range of the Columns in C,

7. each column of B has to have the same number of elements as the number of elements

   in each row of A. There has to be the same number of columns in A as there are rows in B.

III. The Evaluation of
$$c_{jk} = \sum_{i=1}^{n} ( a_{ji} \quad b_{ik} )_{m \times p} \qquad \text{to find the matrix } C_{m \times p}.$$

$c_{11}$ : let j = 1 & k = 1 ;

     $c_{11}$   is the product where i = 1  added to the product where i = 2 and so on to i = n

$c_{12}$ : let j = 1 & k = 2 ;

     $c_{12}$   is the product where i = 1  added to the product where i = 2 and so on to i = n

$c_{13}$ : let j = 1 & k = 3 ;

     $c_{13}$   is the product where i = 1  added to the product where i = 2 and so on to i = n

     and so on  until the j's are covered 1 through m  & the k's are covered 1 through p.

7.16  MATRIX MULTIPLICATION.

I. Introduction.

It has been fairly obvious how to define equality, addition, and product by a scalar when dealing with matrices, but not so with the product of two matrices. Expressions involving products like $a_{34} \cdot b_{34}$ have not occurred in any problem solutions; so this kind of 'product' is of no value to us. Solutions like the following give the impression of multiplication and occur enough times to warrant special study --- this leads into our definition in 7.17

II. Problem:  consider three different coordinate systems in the same plane where one can be changed to another by rotation of axes about a common origin.

1. In a previous mathematics it was proved that a description of a point $P(x_1, x_2)$ in terms of the original $x_1, x_2$ axes can be changed  to the  $P(y_1, y_2)$ format in terms of the rotated $y_1, y_2$ axes by means of the system of equations:

$y_1 = x_1 \cos \theta + x_2 \sin \theta$

$y_2 = x_1 \sin \theta + x_2 \cos \theta$  where $\theta$ is the angle of rotation of the x axes into the y axes,

2. The same kind of thing applies to the $2^{nd}$ rotation changing the $P(y_1, y_2)$ point to a $P(w_1, w_2)$ format involving a  $w_1, w_2$ axes  by means of the system:

$w_1 = y_1 \cos \beta + y_2 \sin \beta$

$w_2 = y_1 \sin \beta + y_2 \cos \beta$  where $\beta$ is the angle of rotation of the y axes into the w axes,

3. We are only interested in the form of the solutions; so can we change the systems to

$$y_1 = x_1 a_{11} + x_2 a_{12} \qquad\qquad w_1 = y_1 b_{11} + y_2 b_{12}$$

$$y_2 = x_1 a_{21} + x_2 a_{22} \qquad\qquad w_2 = y_1 b_{21} + y_2 b_{22}$$

where the a's & b's are constants representing the sines & cosines; this will simplify the mathematics but still describe the rotations.

4. The mathematics involves routine but involved substitutions and simplifications with the subscripts helping keep track of the various components. The final system will be:

$$w_2 = x_1 c_{11} + x_2 c_{12}$$

$$w_2 = x_1 c_{21} + x_2 c_{22}$$

where  $c_{11} = a_{11} b_{11} + a_{12} b_{21}$       and       $c_{12} = a_{11} b_{12} + a_{12} b_{22}$

$c_{21} = a_{21} b_{11} + a_{22} b_{21}$                 $c_{22} = a_{21} b_{12} + a_{22} b_{22}$

5. We see that we can represent each  $c_{jk}$   by

$$c_{jk} \ = \ \sum_{i=1}^{2} ( \ a_{ji} \quad b_{ik} \ )_{2 \times 2}$$

for each element $c_{jk}$ has the form of    $a_{j1} \ b_{1k} \ + \ a_{j2} \ b_{2k}$    ,

6. From the ( $a_{ji} \cdot$  $b_{ik}$  ) in  the $\sum$ we get the impression of Multiplication, and though the work is complex relative to products of real numbers there seems to be a pattern to the operations. We do get the feeling that if we have a matrix A and a matrix B  and operate on them by means of some kind of a matrix multiplication,  • , to obtain a matrix C , the matrix C will solve some of the problems that involve A & B.


III.

1. In the past mathematicians have discovered that the solutions to many different kinds of problems had this pattern; so this pattern is important.

2. In working with this pattern two different graphic devices were discovered to give the linear combinations in 7.16 II.  without all the substitutions and simplifications of  those steps

3. This is so important that we will use one of the devices as the basic definition of Matrix Multiplication. We will study the other later. For convenience we start with 2x2 matrices.

4. We have found no other pattern that has this use; so in showing products of matrices we will use any of the ways that we have had to indicate a product with real numbers.

IV.  As we look at any one of the elements  $c_{jk}$  in II. 4b , we note that the products in the sums seem to be a row of A times a column of B. This leads us naturally into the following definition.

7.17

II. Definition of  the Product of Two   2 x 2  Matrices.

A. The Pattern:

1. Given  $A = \begin{bmatrix} a_{11} & a_{12} \\ a_{21} & a_{22} \end{bmatrix} = (a_{ji})_{2 \times 2}$    &  $B = \begin{bmatrix} b_{11} & b_{12} \\ b_{21} & b_{22} \end{bmatrix} = (b_{ik})_{2 \times 2}$

and   $C = \begin{bmatrix} c_{11} & c_{12} \\ c_{21} & c_{22} \end{bmatrix} = (c_{jk})_{2 \times 2}$                 where

$$c_{jk} \ = \ \sum_{i=1}^{2} ( \ a_{ji} \quad b_{ik} \ )_{2 \times 2}$$

2. Graphic display:

$$A \bullet B = \begin{matrix} \text{1st} \\ \text{2nd} \\ \text{3rd} \\ \text{4th} \end{matrix} \begin{bmatrix} a_{11} & a_{12} \\ \\ a_{21} & a_{22} \end{bmatrix} \bullet \begin{bmatrix} b_{11} & b_{12} \\ \\ b_{21} & b_{22} \end{bmatrix} = \begin{bmatrix} c_{11} & c_{12} \\ c_{21} & c_{22} \end{bmatrix}$$

1st   is the Linear Combination of the 1st row of A with the 1st column of B to get $c_{11}$

2nd   is the Linear Combination of the 1st row of A with the 2nd column of B to get $c_{12}$

3rd   is the Linear Combination of the 2nd row of A with the 1st column of B to get $c_{21}$

4th   is the Linear Combination of the 2nd row of A with the 2nd column of B to get $c_{22}$

3. result:

$$A \bullet B = \begin{bmatrix} [\, a_{11} \quad b_{11} \quad + a_{12} \quad b_{21} \,] & [\, a_{11} \quad b_{12} \quad + a_{12} \quad b_{22} \,] \\ [\, a_{21} \quad b_{11} \quad + a_{22} \quad b_{21} \,] & [\, a_{21} \quad b_{12} \quad + a_{22} \quad b_{22} \,] \end{bmatrix}$$

B. Example: $\begin{bmatrix} 5 & 0 \\ 3 & 2 \end{bmatrix} \begin{bmatrix} 1 & -2 \\ -1 & 0 \end{bmatrix} = \begin{bmatrix} 5 & -10 \\ 1 & -6 \end{bmatrix}$

C. It is clear that:

    1. The number of columns in the first factor had to be the same as the number of rows in the second. Some writers call this being CONFORMABLE with respect to multiplication and use this label for all operations when the orders make the operation possible.

    2. In the final product the numbers of rows are the same as the number of rows in the first factor, and the number of columns are the same as the number of the columns of the second factor.


    III. Definition.   The Product of Two  2 x 2 Matrices in the Short-hand form.

$$(a_{ji})_{2\times 2} \bullet (b_{ik})_{2\times 2} = (c_{jk})_{2\times 2} \quad \text{where} \quad c_{jk} = \sum_{i=1}^{2} (\, a_{ji} \quad b_{ik} \,)_{2 \times 2}$$

some write it as:

$$(a_{ji}) \bullet (b_{ik}) = (c_{jk}) \quad \text{where} \quad c_{jk} = \sum_{i=1}^{2} (\, a_{ji} \quad b_{ik} \,) \quad \text{with} \quad \begin{matrix} j = 1, 2. \\ k = 1, 2. \end{matrix}$$

IV. The Product of Two  2 x 2 Matrices using Vector Dot Products.

1. Given    $A \bullet B \equiv$ $\begin{bmatrix} a_{11} & a_{12} \\ a_{21} & a_{22} \end{bmatrix}$ $\begin{bmatrix} b_{11} & b_{12} \\ b_{21} & b_{22} \end{bmatrix}$

2. think of A as composed of 2 Row matrices [ or vectors]:

$A$ = $\begin{bmatrix} a_1 \\ a_2 \end{bmatrix}$     where    $a_1$ = $\langle a_{11} , a_{12} \rangle$

$a_2$ = $\langle a_{21} , a_{22} \rangle$

3. think of B as composed of 2 Column matrices [ or vectors]:

$B$ = $( b_1 , b_2 )$ where    $b_1$ = $\begin{bmatrix} b_{11} \\ b_{21} \end{bmatrix}$ & $b_2$ = $\begin{bmatrix} b_{12} \\ b_{22} \end{bmatrix}$

4. now   $b_1^T$ = $\langle b_{11} , b_{21} \rangle$ & $b_2^T$ = $\langle b_{12} , b_{22} \rangle$

5. Dot Products:  $a_1 \circ b_1^T = a_{11} b_{11} + a_{12} b_{21}$ ,  $a_1 \circ b_2^T = a_{11} b_{12} + a_{12} b_{22}$

$a_2 \circ b_1^T = a_{21} b_{11} + a_{22} b_{21}$ ,  $a_2 \circ b_2^T = a_{21} b_{12} + a_{22} b_{22}$

6. now $A \bullet B$ = $\begin{bmatrix} \{a_{11} b_{11} + a_{12} b_{21}\} & \{a_{11} b_{12} + a_{12} b_{22}\} \\ \{a_{21} b_{11} + a_{22} b_{21}\} & \{a_{21} b_{12} + a_{22} b_{22}\} \end{bmatrix}$

7. and:   $A \bullet B$ = $\begin{bmatrix} [a_1 \circ b_1^T ] & [ a_1 \circ b_2^T ] \\ [a_2 \circ b_1^T ] & [ a_2 \circ b_2^T ] \end{bmatrix}$     from I.5.

This is sometimes useful in theorical studies. Some writers assume

the b's are transposes without saying so.

8. It is clear how we can expand all of our work from 2 x 2 to m x p matrices.

7.18. Examples.

I.

1. $\begin{bmatrix} 2 & 1 \\ 3 & 4 \end{bmatrix} \begin{bmatrix} 1 & 0 \\ -2 & -1 \end{bmatrix} = \begin{bmatrix} 0 & -1 \\ -5 & -4 \end{bmatrix}$

2. $\begin{bmatrix} 3 & 2 & -1 \\ 0 & 4 & 6 \end{bmatrix} \begin{bmatrix} 1 & 0 & 2 \\ 5 & 3 & 1 \\ 6 & 4 & 2 \end{bmatrix} = \begin{bmatrix} 7 & 2 & 6 \\ 56 & 36 & 16 \end{bmatrix}$

II.

1. $\begin{bmatrix} 1 & 0 & 3 \end{bmatrix} \begin{bmatrix} 1 & 0 \\ 0 & 1 \\ -1 & 2 \end{bmatrix} = \begin{bmatrix} -2 & 6 \end{bmatrix}$   note  a   1 x 3   times a   3 x 2

⊢————————————→
OK

2. but $\begin{bmatrix} 1 & 0 \\ 0 & 1 \\ -1 & 2 \end{bmatrix} \begin{bmatrix} 1 & 0 & 3 \end{bmatrix}$

can not be multiplied because a   3 x 2   times a   1 x 3

⊢————————→
NO

3. $\begin{bmatrix} 1 & 2 \\ 0 & -1 \end{bmatrix} \begin{bmatrix} 0 & 1 \\ -1 & -2 \end{bmatrix} = \begin{bmatrix} -2 & -3 \\ 1 & 2 \end{bmatrix}$

but $\begin{bmatrix} 0 & 1 \\ -1 & -2 \end{bmatrix} \begin{bmatrix} 1 & 2 \\ 0 & -1 \end{bmatrix} = \begin{bmatrix} 0 & -1 \\ -1 & 0 \end{bmatrix}$   not the same

4. So in general either  A B  ≠  B A    or  B A  can not be done; there may be special cases

where  A B  =  B A.

5. Therefore the Product of Matrices is Not Commutative.

6. Note the subscripts :  $C_{pi}$   [ $A_{ij}$   $B_{jk}$ ]   =  $C_{pi}$   $D_{ik}$   =  $E_{pk}$

The middle index is eliminated after each Product.


7.19 A Different Graphic Method to find Matrix Products.

1.   If   A  =  $\begin{bmatrix} 3 & 2 & -1 \\ 0 & 4 & 6 \end{bmatrix}$   &   B  =  $\begin{bmatrix} 1 & 0 & 2 \\ 5 & 3 & 1 \\ 6 & 4 & 2 \end{bmatrix}$   ,  then  find  A B .

2.

|   |   |   | 1 | 0 | 2 |
|---|---|---|---|---|---|
|   | B | ↓ | 5 | 3 | 1 |
|   | → |   | 6 | 4 | 2 |
| A | 3 | 2 | -1 | 7 | 2 | 6 |
|   | 0 | 4 | 6 | 56 | 36 | 16 |

A B

[3 ][1 ] + [2 ][5 ] + [-1][6 ] =   7

[3 ][0 ] + [2 ][3 ] + [-1][4 ] =   2

[3 ][2 ] + [2 ][1 ] + [-1][  ] =   6

[0 ][1 ] + [4 ][5 ] + [6 ][6 ] =  56

[0 ][0 ] + [4 ][3 ] + [6 ][4 ] =  36

[0 ][2 ] + [4 ][1 ] + [6 ][2 ] =  16

note: there is a special method to check the answer, but it is simpler to just

double-check your work.

3. Example: Given the A & B in #1 and C = ( -1   0 ); find   C ( A B )

|   |   |   |   | 1 | 0 | 2 |
|---|---|---|---|---|---|---|
|   | B |   |   | 5 | 3 | 1 |
|   |   |   |   | 6 | 4 | 2 |
| A | 3 | 2 | -1 | 7 | 2 | 6 |
|   | 0 | 4 | 6 | 56 | 36 | 16 | A B |
| C |   | -1 | 0 | -7 | -2 | -6 | C ( A B ) |

7.20. PROPERTIES of MATRIX PRODUCTS.

Note: the proofs involve sums and products of real numbers ; so in addition to what was said before about the Sigma notation the laws of real numbers apply through out the proofs.

Review 7.6 - 7.9.

I.    ASSOCIATIVE LAW with a Scalar, s , and Two Matrices.

Given:  $A = ( a_{ji} )_{m \times n}$        &      $B = ( b_{ik} )_{n \times p}$      and     's'     a scalar,

We see that: $i = 1, 2, 3, ..., n$ ;   $j = 1, 2, 3, ..., m$ ;   $k = 1, 2, 3, ..., p$,

Then: $s [ A B ] = [ s A ] B = A [ s B ] = s A B = A s B = A B s$

Proof

1.
$$s [ A B ] = s [ \sum_{i=1}^{n} ( a_{ji} \ b_{ik} )_{m \times p} ]$$

$$= \sum_{i=1}^{n} s ( a_{ji} \ b_{ik} )_{m \times p} \qquad : s \text{ is not affected by } \sum$$

either:  2. $s [ A B ] = \sum_{i=1}^{n} ( [ s a]_{ji} \ b_{ik} )_{m \times p} = [ s A ] B$

or:  3. $s [ A B ] = \sum_{i=1}^{n} ( a_{ji} \ [s b]_{ik} )_{m \times p} = A [ s B ]$

4. the symbols of grouping are not necessary; so

$$s [ A B ] = [ s A ] B = A [ s B ] = s A B = A s B = A B s$$

II.   ASSOCIATIVE LAW with Three Matrices.

Given:   $A = (a_{ji})_{m \times n}$      &      $B = (b_{ik})_{n \times p}$  & $C = (c_{kh})_{p \times r}$

We note that: $i = 1, 2, 3, \ldots, n$ ;    $j = 1, 2, 3, \ldots, m$ ;

$k = 1, 2, 3, \ldots, p,$      $h = 1, 2, 3, \ldots, r$

OR we could write this as:

Given:   $A = (a_{ji})$      &      $B = (b_{ik})$   & $C = (c_{kh})$

where:       $i = 1, 2, 3, \ldots, n$ ;    $j = 1, 2, 3, \ldots, m$ ;

$k = 1, 2, 3, \ldots, p,$      $h = 1, 2, 3, \ldots, r$

Then:   $A [BC] = [AB] C = ABC.$

Proof using the first given form:

1. $A [BC] = [(a_{ji})_{m \times n}][\sum\limits_{k=1}^{p} (b_{ik} c_{kh})_{n \times r}]$

$= \sum\limits_{i=1}^{n} (a_{ji} \sum\limits_{k=1}^{p} b_{ik} c_{kh})_{m \times r}$

$= \sum\limits_{i=1}^{n} \sum\limits_{k=1}^{p} (a_{ji} [b_{ik} c_{kh}])_{m \times r}$

by a property of the Sigma notation -- the second $\sum$ is a function of $k$; so the a's behave like constants,

$= \sum\limits_{k=1}^{p} (\sum\limits_{i=1}^{n} [a_{ji} b_{ik}]_{m \times p} c_{kh})_{m \times r}$

by The  Associative Law for a, and in a finite series the sigmas can be reversed,

$= \sum\limits_{k=1}^{p} ([a_{ji} b_{ik}]_{m \times p} c_{kh})_{m \times r}$

$= [AB] C$

2. So: $A [BC] = [AB] C$   thus the symbols of grouping are not necessary; so

$= ABC.$          So, proved.

III. Exercise. Given

$$A \begin{bmatrix} 1 & 1 \\ 1 & -1 \end{bmatrix}, \quad B \begin{bmatrix} 0 & 1 \\ -1 & 1 \end{bmatrix}, \quad C \begin{bmatrix} -1 & 1 \\ 1 & 0 \end{bmatrix}$$

Find: A [ B C ] and [ A B ] C.

## 7.21 THE DISTRIBUTIVE LAWS.

Given: $A = (a_{ji})$ & $B = (b_{ji})$ & $C = (c_{ik})$

Where in all cases $i = 1,2,3,\ldots,n$ ; $j = 1,2,3,\ldots,n$ ; $k = 1,2,3,\ldots,p$,

and $\sum$ implies $i = 1,2,3,\ldots,n$

Note: we are using the other way to keep track of the ranges of the indicies.

I.    Prove: [ A + B ] C = A C + B C.

[ A + B ] C $= [(a_{ji}) + (b_{ji})] [(c_{ik})] = [(a+b)_{ji}][(c_{ik})]$

$= \sum([a+b]_{ji} c_{ik}) = \sum([a+b]c)_{jk} = \sum\{(c+bc)_{jk}\}$

$= \sum (ac)_{jk} + \sum (bc)_{jk} = \sum (a_{ji} c_{ik}) + \sum (b_{ji} c_{ik})$

$= A C + B C$

II. Theorem.    Given: a matrix $A = (a_{ij})_{m \times n}$ and the matrix $X = (x_{jk})_{n \times p}$ ,

let a function f be such that $f(X) = A X$ ;

Prove that f is Linear.

1. $f(X_1 + X_2) = A (X_1 + X_2) = A X_1 + A X_2 = f(X_1) + f(X_2)$

note: $f(kX) = A (kX) = k (AX) = k f(X)$

2. so   f is Linear.

III. Exercises:

1. Prove:    C ( A + B ) = C A + C B       using the above givens.

2. Find A B where $A = \begin{bmatrix} 1 & 1 \\ 2 & 2 \end{bmatrix}$ & $B = \begin{bmatrix} -1 & 1 \\ 1 & -1 \end{bmatrix}$

7.22. Theorem: If $A = (a_{ji})_{m \times n}$ & $B = (b_{ik})_{n \times p}$ ,

Then $[A B]^T = B^T A^T$

Proof:

1. for $j = 1, 2, ..., m$ & $i = 1, 2, ..., n$ & $k = 1, 2, ..., p$

$$A B = \sum_{i=1}^{n} a_{ji} \, b_{ik}$$

2. $[A B]^T = \sum_{i=1}^{n} a_{ij} \, b_{ki} = \sum_{i=1}^{n} b_{ki} \, a_{ij}$

3. $B^T = (b_{ki})$ & $A^T = (a_{ij})$

4. $B^T A^T = \sum_{i=1}^{n} (b_{ki} \, a_{ij})$

5. Then $[A B]^T = B^T A^T$

## 7.23 The INVERSE of a SQUARE MATIRX.

1. In the real number system products lead to division, and that is handled by starting with Reciprocals and Inverses. As: if a b = 1 then b = 1/a where a ≠ 0 . 1/a is the Reciprocal of a , and this is sometimes written as : $b = a^{-1}$ . We can also write the equation a b = 1 in the form of : $a = 1/b = b^{-1}$ where b ≠ 0 . In matrices we run into trouble if we try this concept with non-square matrices. Also, we do not use the reciprocal form 1/A with matrices because it is better to think of the division concept as a product of the inverse form as in $A/B = A \bullet B^{-1} = A \, B^{-1}$ .

2. Definition of the INVERSE of a Square Matrix.

Given square matrices $A_{nn}$ & $B_{nn}$ where $A_n B_n = I_n$ and $B_n A_n = I$, & A,B ≠ 0, then

$B_n$ is the Inverse of A and written as $A^{-1}$ ;

also $A_n$ is the Inverse of B and written as $B^{-1}$ .

note: $A_{n \times n} = A_{nn} = A_n$ ; all of these symbols mean the same thing.

3. Example: Let $A = \begin{bmatrix} 1 & 4 \\ 2 & 9 \end{bmatrix}$ and $B = \begin{bmatrix} 9 & -4 \\ -2 & 1 \end{bmatrix}$

then $AB = \begin{bmatrix} 1 & 4 \\ 2 & 9 \end{bmatrix} \begin{bmatrix} 9 & -4 \\ -2 & 1 \end{bmatrix} = \begin{bmatrix} 1 & 0 \\ 0 & 1 \end{bmatrix} = I_2$

and $BA = \begin{bmatrix} 9 & -4 \\ -2 & 1 \end{bmatrix} \begin{bmatrix} 1 & 4 \\ 2 & 9 \end{bmatrix} = \begin{bmatrix} 1 & 0 \\ 0 & 1 \end{bmatrix} = I_2$

so $A^{-1} = \begin{bmatrix} 9 & -4 \\ -2 & 1 \end{bmatrix}$    and $B^{-1} = \begin{bmatrix} 1 & 4 \\ 2 & 9 \end{bmatrix}$

4. Find $A^{-1}$    if $A = \begin{bmatrix} 1 & -2 \\ -3 & -1 \end{bmatrix}$

$$\begin{bmatrix} 1 & -2 \\ -3 & -1 \end{bmatrix} \begin{bmatrix} a & b \\ c & d \end{bmatrix} = \begin{bmatrix} 1 & 0 \\ 0 & 1 \end{bmatrix}$$

$$\left\{ \begin{array}{l} a - 2c = 1 \\ -3a - c = 0 \end{array} \right\} \quad \& \quad \left\{ \begin{array}{l} b - 2d = 0 \\ -3b - d = 1 \end{array} \right\}$$

thus $c = -3/7$ & $a = 1/7$    and $b = -2/7$ & $d = -1/7$

so $A^{-1} = \begin{bmatrix} 1/7 & -2/7 \\ -3/7 & -1/7 \end{bmatrix}$

5. as an exercise check the product to see if it is I   .


7.24  The discovery of additional matrix concepts is helped by additional study of Systems of Linear Equations. Our prime consideration here is that Linear refers to First Degree Equations. Because this text is restricted to such equations, we will in many cases refer to just "Systems of Equations". This extension of the study of mtrices is involved enough to warrant whole chapters of study.

8.0 Introduction.

One way that mathematicians have developed to solve the problems of the real world
is the represent the unknowns as variables like $x_1$ , $x_2$ ,...,$x_n$ , and then find
true equation(s) involving these x's. The x's usually end up in a system of such
equations; a system to be solved. We will now study systems that are of the first
degree. Because of our previous work we use the label Linear System of Equations.

We have mentioned that one of the first uses of Matrices was to help solve such
systems. The student has had some experience in dealing with such systems in
previous studies and has seen some value in this kind of study. A further study of
solutions of these systems will increase his knowledge of matrices as well . It is
well here to first review what is known about such systems and then to add to this
knowledge where appropriate.

8.1 A System of m First Degree Equations in n unknowns -- usually called a

system of "Simultaneous Linear Equations" because each equation's left side

is a Linear Function; so they plot as Straight Lines. As an Exercise prove this.

1. as in:
$$\begin{cases} a_{11} \ x_1 + a_{12} \ x_2 + \ldots + a_{1n} \ x_n = b_1 \\ a_{21} \ x_1 + a_{22} \ x_2 + \ldots + a_{2n} \ x_n = b_2 \\ \ldots\ldots\ldots\ldots\ldots\ldots\ldots\ldots\ldots\ldots\ldots\ldots\ldots \\ \ldots\ldots\ldots\ldots\ldots\ldots\ldots\ldots\ldots\ldots\ldots\ldots\ldots \\ a_{m1} \ x_1 + a_{m2} \ x_2 + \ldots + a_{mn} \ x_n = b_m \end{cases}$$

2. is called a system because each equation is part of a Whole; even though the
word Linear as a Line makes sense in only one and two unknowns, the label linear is used
for all numbers of unknowns;

3. and the x's are the Unknowns to be evaluated, the a's are the Coefficients which
are given Constants or Scalars, the b's are the Constants of the System.

4. if all the b's are 0, then the system is called Homogeneous; if at least one of
the b's is not zero, then it is called Non-Homogeneous,

5. a Solution to the system is the Set of Numbers $x_1$ , $x_2$ ,...,$x_n$ that satisfy
all m equations; that is: on substitution each equation simplifies to an Identity,

6. if the system is Homogeneous , it has at least the Trivial Solution of

$x_1 = x_2 = \ldots = x_n = 0$  --- it may have an Infinite Set of Other Values.

1. We can write it in a form that uses Matrices ; as :

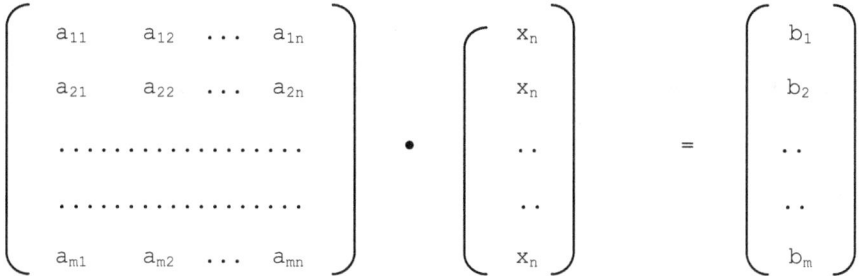

2. then the Matrix Equation of the system is:

$$A \qquad \bullet \qquad X \qquad = \qquad B$$

or in the short-hand form of:

$$(a_{ji})_{m \times n} \qquad \bullet \qquad (x_{ij})_{n \times 1} \qquad = \qquad (b_{ji})_{m \times 1}$$

3. the Augmented Matrix of the system is a matrix of all of the Constants arranged in the given format; it is usually labeled  B  . It actually represents the given system. So, it looks like:

$$B \;=\; \begin{pmatrix} a_{11} & a_{12} & \cdots & a_{1n} & b_1 \\ a_{21} & a_{22} & \cdots & a_{2n} & b_2 \\ \cdots & \cdots & \cdots & \cdots & \cdots \\ \cdots & \cdots & \cdots & \cdots & \cdots \\ a_{m1} & a_{m2} & \cdots & a_{mn} & b_m \end{pmatrix}$$

8.3 Elementary Operations on Matrices.

I.

1. In order to solve this kind of system by matrices  we need to study certain special operations that can be performed on matrices. These change the form of a matrix in order to accomplish a definite purpose or to display a different set of information; a matrix has no value so we are not involved with "values"  but just form.

2. Definition. We perform "An Elementary Row Operation" on a matrix by a finite application of the following operations  --- either done singly  or in some combination of the any of the three:

      a. Either Interchange any two Rows any number of times,

      b. or multiply any Row by a non-zero constant [scalar],

      c. or multiply the ith row by a  scalar and then add this new row to

         the jth row to get a new jth row:  i ≠ j without changing the ith row.

   3. If there is a good reason for it , we can do the same thing all the way through

         with columns,

II. If we change a matrix A by Elementary Row Operations to a final matrix

         A' , we say that the original and the new matrix are Row Equivalent

         [there is no row Equal unless the two forms are identical].


   8.4. The equivalency of elementary row operations.

   I.

   1.  A question naturally arises at this time; is this the same mathematical

 equivalency as in 2.1? Or, is it a word that is used loosely to describe some concept

that is similar in nature to our previous use? Can we apply some combination of the

 operations when changing A into A' that sets up a relation that satisfies the three

tests of an equivalent class? Note: with each test for equivalency we have to test three

tests for the three row operations.

   2. For convenience of writing use ~ to represent the row operations as we answer that

question. We do not wish to get too deep into this theory, but if the student will look

at specific examples it would be clear why the conclusions below are correct. Normally

 we envision the equivalence definition in terms of  a. A ~ A , b. if A ~ B, then

 B ~ A, and c. if A ~ B , & B ~ C , then A ~ C. Using A' for B & A'' for C is better here

for A' not only gives a symbol for the second matrix but tells us how we got it –the same

for using A'' instead of C.

   II.

   1. A ~ A  --the first test: can we perform elementary operations and get  A from A?

      a. in A interchange the ith & jth rows and then interchange the same ith & jth rows

to get A again; so A ~ A;

      b. multiply any row by a constant k and then multiply that same row by 1/k which

again changes A back into A; so A ~ A.

      c. mentally multiply an ith row by a constant k and add it to an kth row changing only that jth row. Now mentally multiply the ith  row by -k and add it to this changed

kth row. . This gives us the original kth row. A changes back to A;  so  A ~ A.

   2. If  A ~ A', then A' ~ A  -- the second test.

      a. Assume the ith & jth rows of A were interchanged in A to get the A' in the

first place. Now interchange the ith & jth rows of A' to get A''; but this gives us A

again -- or A'' = A . So, if A ~ A', then A' ~ A.

      b. Assume the ith row of A was multiplied by k to get the A'. Now multiply  the

ith row of A' by 1/k to get A''; we note that this A'' = A. So, again if A ~ A', then

A' ~ A.

      c. Assume the ith row of A was multiplied by k and added to the jth row

 to get A'. Now  multiply the ith row of A' by -k and add to the jth row to get A'';

 this A'' = A. So, again A' ~ A when A ~ A'.

   3. If A ~ A' and A' ~ A'' , then A ~ A''   -- the third test. Note that we have

really proved this in step # 4. In # 4 we were using the equalities to get a relation

between A & A'. For step # 5 we can use the same equalities to show the relation

between A & A''. We note that in all the proofs we did something to a matrix and  then

did the inverse operation to get back to A .

III. We have proved that performing the Elementary Operations on a matrix A gives us a

matrix A' that is Equivalent to A. We will see what this means later. Now we will study

more matrix theory.


8.5. Elementary Operation Matrices.

 I. Introduction.

      1.One problem with changing the form of a matrix by using the elementary operations

is that we are following a set of instructions of what to do. We are not putting this

down in a mathematical form like creating  a set of equations to be solved. However, the

instructions do change one form of a matrix into another; so we might accomplish this by

multiplying the given matrix by another special matrix.

      2. Labeling the special matrix, E, we will try to do this with an example.

 II. Given the matrix A = $\begin{bmatrix} -1 & 3 \\ 2 & 4 \end{bmatrix}$

1. find a matrix $E_1$ so that $E_1$ A interchanges the two rows:

a. Use the form: $\begin{bmatrix} w & x \\ y & z \end{bmatrix} \begin{bmatrix} -1 & 3 \\ 2 & 4 \end{bmatrix} = \begin{bmatrix} 2 & 4 \\ -1 & 3 \end{bmatrix}$ which gives us:

$\begin{bmatrix} [-w+2x] & [3w+4x] \\ [-y+2z] & [3y+4z] \end{bmatrix} = \begin{bmatrix} 2 & 4 \\ -1 & 3 \end{bmatrix}$ thus $\left\{ \begin{array}{l} -w + 2x = 2 \\ 3w + 4x = 4 \end{array} \right\}$ & $\left\{ \begin{array}{l} -y + 2z = -1 \\ 3y + 4z = 3 \end{array} \right\}$

or $\left\{ \begin{array}{l} 2w - 4x = -4 \\ 3w + 4x = 4 \end{array} \right\}$ and $\left\{ \begin{array}{l} 2y - 4z = 2 \\ 3y + 4z = 3 \end{array} \right\}$

adding:      $5w = 0$      &      $5y = 5$

              $w = 0$      &       $y = 1$

and          $2x = 2$      &   $-1 + 2z = -1$

or           $x = 1$       &       $z = 0$

to interchange the two rows of this A:   $E_1 = \begin{bmatrix} 0 & 1 \\ 1 & 0 \end{bmatrix}$

b. check: $\begin{bmatrix} 0 & 1 \\ 1 & 0 \end{bmatrix} \begin{bmatrix} -1 & 3 \\ 2 & 4 \end{bmatrix}$ does equal $\begin{bmatrix} 2 & 4 \\ -1 & 3 \end{bmatrix}$ label this A' for this example,

c. Therefore for each matrix A we have a Method to find a matrix $E_1$ such that the product $E_1$ A interchanges Two Rows.

d. Sometimes things are so simple that one can guess the numbers and their placement in $E_1$ , but the method in step #a. will always work.

e. Note this $E_1$ is a reverse of the $I_n$ matrix; so try this idea on a 3x3 matrix:

$\begin{bmatrix} 0 & 0 & 1 \\ 0 & 1 & 0 \\ 1 & 0 & 0 \end{bmatrix} \begin{bmatrix} a & b & c \\ d & e & f \\ g & h & k \end{bmatrix} = \begin{bmatrix} g & h & k \\ d & e & f \\ a & b & c \end{bmatrix}$ which does interchange rows 1 & 3.

f. The student might find it interesting to use other forms where each row has 0's and one 1 in different combinations to interchange two rows on different sized matrices.

2. Find matrix $E_2$ so that $E_2$ A' multiplies a row by a constant:

a. multiply 1st row , say, by 3: $\begin{bmatrix} w & x \\ y & z \end{bmatrix}\begin{bmatrix} 2 & 4 \\ -1 & 3 \end{bmatrix}$ must equal $\begin{bmatrix} 6 & 12 \\ -1 & 3 \end{bmatrix}$

b. $\begin{Bmatrix} 2w - x = 6 \\ 4w + 3x = 12 \end{Bmatrix}$ & $\begin{Bmatrix} 2y - z = -1 \\ 4y + 3z = 3 \end{Bmatrix}$ implies $\begin{Bmatrix} 6w - 3x = 18 \\ 4w + 3x = 12 \end{Bmatrix}$ & $\begin{Bmatrix} 6y - 3z = -3 \\ 4y + 3z = 3 \end{Bmatrix}$

c. and $10w = 30$ & $10y = 0$ or $w = 3$ & $y = 0$

then $x = 0$      $z = 1$

d. so $E_2$ = $\begin{bmatrix} 3 & 0 \\ 0 & 1 \end{bmatrix}$ check $\begin{bmatrix} 3 & 0 \\ 0 & 1 \end{bmatrix}\begin{bmatrix} 2 & 4 \\ -1 & 3 \end{bmatrix}$ does equal $\begin{bmatrix} 6 & 12 \\ -1 & 3 \end{bmatrix}$ = A''

e. note if we let $E_2'$ = $\begin{bmatrix} 1 & 0 \\ 0 & k \end{bmatrix}$ then $\begin{bmatrix} 1 & 0 \\ 0 & k \end{bmatrix}\begin{bmatrix} 2 & 4 \\ -1 & 3 \end{bmatrix}$ = $\begin{bmatrix} 2 & 4 \\ -k & 3k \end{bmatrix}$ 2nd row multiplied by k

f. Therefore for each matrix A we have a Method to find a matrix $E_2$ such that the product $E_2$ A multiplies a row by a constant k.

3. Find matrix $E_3$ so that $E_3$ A'' changes the jth row by mentally multiplying the ith row by a constant c and actually adding it to the jth row to get the new jth row.

a. multiply the 2nd row , say, by -2 and add it, say, to the 1st row :

$\begin{bmatrix} w & x \\ y & z \end{bmatrix}\begin{bmatrix} 6 & 12 \\ -1 & 3 \end{bmatrix}$ = $\begin{bmatrix} 8 & 6 \\ -1 & 3 \end{bmatrix}$

b. $\begin{Bmatrix} 6w - x = 8 \\ 12w + 3x = 6 \end{Bmatrix}$ & $\begin{Bmatrix} 6y - z = -1 \\ 12y + 3z = 3 \end{Bmatrix}$ or $\begin{Bmatrix} 18w - 3x = 24 \\ 12w + 3x = 6 \end{Bmatrix}$ & $\begin{Bmatrix} 18y - 3z = -3 \\ 12y + 3z = 3 \end{Bmatrix}$

c.      $30w = 30$          $30y = 0$

        $w = 1$            $y = 0$

d.         $x = -2$           $z = 1$

e.   $E_3 = \begin{bmatrix} 1 & -2 \\ 0 & 1 \end{bmatrix}$ ;   $\begin{bmatrix} 1 & -2 \\ 0 & 1 \end{bmatrix} \begin{bmatrix} 6 & 12 \\ -1 & 3 \end{bmatrix} = \begin{bmatrix} 8 & 6 \\ -1 & 3 \end{bmatrix}$   $= \quad A'''$

f. Therefore for each matrix A we have a method to find a matrix $E_3$ such that the product $E_3$ A multiplies the ith row by a constant c and adds it to the jth row to change the jth row.

III. Do all of the operations in the example in part II. to develop a 'master' E:

A. Given $A = \begin{bmatrix} -1 & 3 \\ 2 & 4 \end{bmatrix}$ we found

$E_1 = \begin{bmatrix} 0 & 1 \\ 1 & 0 \end{bmatrix}$ & $E_2 = \begin{bmatrix} 3 & 0 \\ 0 & 1 \end{bmatrix}$ & $E_3 = \begin{bmatrix} 1 & -2 \\ 0 & 1 \end{bmatrix}$

where : First, we interchanged the two rows to get A' ; second, we multiplied the 1$^{st}$ row of A' by 3 to get A''; finally, we multiplied the 2$^{nd}$ row of A'' by -2 and added it to the 1$^{st}$ row to get A'''; or as we saw:

1.   $E_1$   A $= \begin{bmatrix} 2 & 4 \\ -1 & 3 \end{bmatrix} = A'$

2. $E_2$   A' $= \begin{bmatrix} 6 & 12 \\ -1 & 3 \end{bmatrix} = A''$

3. $E_3$   A'' $= \begin{bmatrix} 8 & 6 \\ -1 & 3 \end{bmatrix} = A'''$

B.

1. What we did was :   $E_3$   { $E_2$   [ $E_1$   A ] } $= A'''$ ,

2. it looks like we could find an  $E = E_3$ {$E_2$   [$E_1$ ] } such that  E  A  $= A'''$ ,

3. or $\underset{E_2}{\begin{bmatrix} 3 & 0 \\ 0 & 1 \end{bmatrix}} \underset{E_1}{\begin{bmatrix} 0 & 1 \\ 1 & 0 \end{bmatrix}} = \begin{bmatrix} 0 & 3 \\ 1 & 0 \end{bmatrix}$ and then $\begin{bmatrix} 1 & -2 \\ 0 & 1 \end{bmatrix} \underset{E_3}{\begin{bmatrix} 0 & 3 \\ 1 & 0 \end{bmatrix}} = \begin{bmatrix} -2 & 3 \\ 1 & 0 \end{bmatrix} = E$

so $\begin{bmatrix} -2 & 3 \\ 1 & 0 \end{bmatrix} \begin{bmatrix} -1 & 3 \\ 2 & 4 \end{bmatrix} = \begin{bmatrix} 8 & 6 \\ -1 & 3 \end{bmatrix}$   which is actually   A'''

IV. We have shown a Method of finding a matrix E such that the product  E A  gives the
same result as performing a series of row operations on A; however in  calculations it is
normally easier to do the series of operations instead of doing the work  that is
necessary to find the master matrix  E .  However, in a lot of theory work we will use
this product to represent the operations.

8.6 Echelon Form of a Matrix.

   I.    1. We normally will be using the elementary operations to change the form of a
matrix  to a "simplest" form called the Echelon form; so the echelon  form of a matrix
is equivalent to the original matrix. The value of this form is obvious from its
structure as we will see below. This is one of the forms that has not been standardized;
we will use the one below. Note, however, a reduced echelon form that appears later.

      2. The Echelon form of a matrix ( $c_{ij}$ )$_{n \times m}$    is one that can be represented by

         a. $c_{ij}$ = 0 for  i > j   [note: some of the other elements also might be zero],

         b. the first non-zero element in a row is 1 ,

         c. all zero rows are at the bottom of the echelon form.

3. as:
$$\begin{bmatrix} 1 & c_{12} & c_{13} \\ 0 & 1 & c_{23} \\ 0 & 0 & 1 \end{bmatrix} \text{ \& } \begin{bmatrix} 1 & c_{12} & c_{13} & c_{14} & c_{15} \\ 0 & 1 & c_{23} & c_{24} & c_{25} \\ 0 & 0 & 1 & c_{34} & c_{35} \\ 0 & 0 & 0 & 1 & c_{45} \end{bmatrix} \text{ \& } \begin{bmatrix} 1 & 6 & 7 & 8 \\ 0 & 1 & 2 & 1 \\ 0 & 0 & 1 & 9 \\ 0 & 0 & 0 & 1 \end{bmatrix}$$

$$\text{\& } \begin{bmatrix} 1 & c_{12} & c_{13} & c_{14} \\ 0 & 1 & c_{23} & c_{24} \\ 0 & 0 & 1 & c_{34} \\ 0 & 0 & 0 & 1 \\ 0 & 0 & 0 & 0 \end{bmatrix} \text{ \& } \begin{bmatrix} 1 & c_{12} & c_{13} & c_{14} \\ 0 & 1 & c_{23} & c_{24} \\ 0 & 0 & 0 & 1 \end{bmatrix}$$

note this
partitioning
with zero's

   II.  We can change a matrix to the echelon form by:

      1.

         a. hunt for  a row whose first element is not zero; interchange rows so that
this row is now the first row – a row whose first c = 1 is best, or if none then do b. ,

b. if necessary divide this row by the number that makes $c_{11} = 1$ ; this move is

not mathematical necessary for equivalency, but it simplifies all work and the use of the

concept so much that it is a Must  ---this final first row is called the Pivotal Row,

c.mentally multiply this new first row by a number that makes  $c_{21} = 0$ when

mentally added to the second row  -- actually write in the addition as the new 2$^{nd}$ row,

  d. again do this same thing to the first row so that in the third row the $c_{31} = 0$,

  e. continue this procedure so that the c's in the rest of the first column

 are all zero;

  2.

  a. Do the same as in step # 1, but we will be working with the second row and down

the second column ,

   b. divide 2$^{nd}$ row by a number that makes  $c_{22} = 1$ ,

  c. do the process in 1.c - e. until all of the other  c's in the 2$^{nd}$ column are zero;

 3. continue this process down the rows until there is no row left to add to; or until

all  $c_{ij} = 0$ for i > j. We can always get an echelon form no matter the m or n.

  III. Example : find the echelon form of this given matrix:

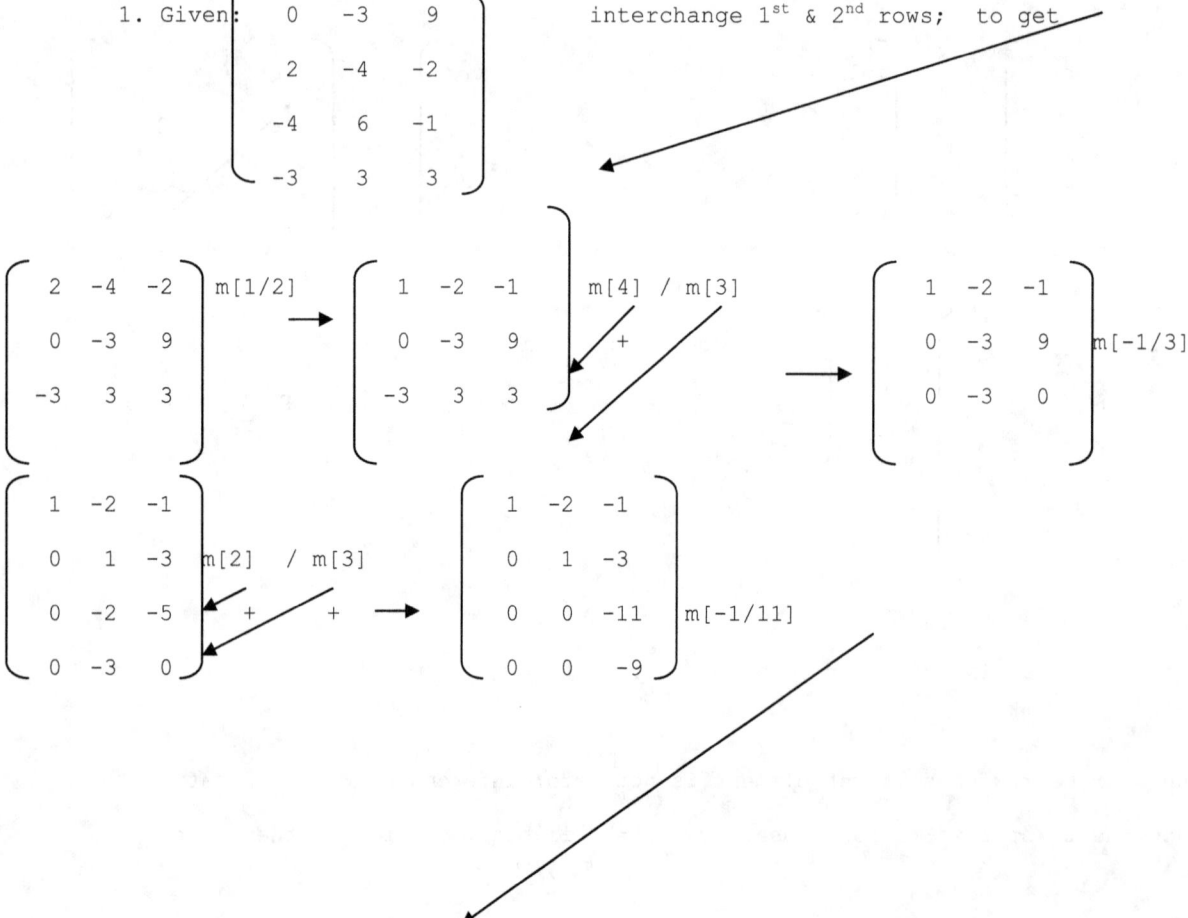

$$
\begin{bmatrix} 1 & -2 & -1 \\ 0 & 1 & -3 \\ 0 & 0 & 1 \\ 0 & 0 & -9 \end{bmatrix} \begin{array}{c} \\ \\ m[9] \\ + \end{array} \longrightarrow \begin{bmatrix} 1 & -2 & -1 \\ 0 & 1 & -3 \\ 0 & 0 & 1 \\ 0 & 0 & 0 \end{bmatrix} \qquad \text{the Echelon form}
$$

IV. The Reduced Echelon Form of  A  .

   1. When the 1's that we worked so hard to get in the echelon form create a

diagonal [a straight line]; we call it The Principal Diagonal. Sometimes the 1's do not

form a straight line, but they always do when the  matrix is square. If A is a square

matrix , then this form is called an Upper Triangular Matrix.

   2. We worked as hard with the operations to get 0's below this diagonal. In addition

we can apply the operations on the rows to get 0's above the diagonal.  This is called

the Reduced Echelon Form ; we will use this form sometimes. If A is square matrix, then

this reduced form is the Unit [ or Identity ] matrix, $I_n$   .

   3. As in this example:

$$
\begin{bmatrix} 1 & 0 & -4 \\ 3 & 1 & -8 \\ 5 & 2 & -10 \end{bmatrix} \begin{array}{c} m[-3] \ \ / \ m[-5] \\ + \\ + \end{array} \longrightarrow \begin{bmatrix} 1 & 0 & -4 \\ 0 & 1 & 4 \\ 0 & 2 & 10 \end{bmatrix} \begin{array}{c} \\ m[-2] \\ + \end{array}
$$

$$
\begin{bmatrix} 1 & 0 & -4 \\ 0 & 1 & 4 \\ 0 & 0 & 2 \end{bmatrix} \begin{array}{c} \\ \\ m[1/2] \end{array} \longrightarrow \begin{bmatrix} 1 & 0 & -4 \\ 0 & 1 & 4 \\ 0 & 0 & 1 \end{bmatrix} \qquad \begin{array}{l} \text{the} \\ \text{Echelon} \\ \text{form} \end{array}
$$

$$
\begin{bmatrix} 1 & 0 & -4 \\ 0 & 1 & 4 \\ 0 & 0 & 1 \end{bmatrix} \begin{array}{c} m[1] \\ \\ \end{array} \longrightarrow \begin{bmatrix} 1 & 0 & -4 \\ 0 & 1 & 0 \\ 0 & 0 & 1 \end{bmatrix} \begin{array}{c} \\ \\ m[4] \end{array} \begin{bmatrix} 1 & 0 & 0 \\ 0 & 1 & 0 \\ 0 & 0 & 1 \end{bmatrix} \ = \ I_3 \ , \quad \begin{array}{l} \text{the Unit or Identity} \\ \text{Matrix; thus the} \\ \text{Reduced Echelon Form.} \end{array}
$$

8.7 Rank of a Matrix.

I. Introduction.

1. In general when solving systems of equations we may get a single set of numbers as the solution, or no solution , or many sets as solutions;

2.  a. the system :  $\begin{Bmatrix} x + y = 5 \\ x + y = 1 \end{Bmatrix}$  obviously has no solution, because subtracting we get $0 = 4$ which is not true;

   b. while the system: $\begin{Bmatrix} x + y = 5 \\ 2x + 2y = 10 \end{Bmatrix}$ has solutions: $(1,4),(2,3),( 10, -5), \dots.$

   c. the system $\begin{Bmatrix} x + y = 5 \\ x - y = 1 \end{Bmatrix}$ on adding gives us $2x = 6$ or $x = 3$ and thus $y = 2$ for

      the only solution of $(3,2)$.

3. the question naturally arises of how can we tell that there exist a solution and that the solution is unique?  In many of our cases we wish to answer this question before the computations to get a solution -especially if the system is of a large order where the computations could be of some length.

4. the concept of the Rank of a Matrix is used to do that.


II. Definition. The Row Rank of a matrix $A = ( a_{ij} )_{n \times m}$ ,is the Maximum Number of Linearly Independent Row Matrices [or Row Vectors]--- symbolized by "rank A" or "r". Review section 4.1. on the concept of linearly independent.

III. Given: matrix $A = ( a_{ij} )_{m \times n}$ :

1. To find the rank r we study:

$k_1 <a_{11} ,a_{12} ,\dots,a_{1n} > + k_2 <a_{21} ,a_{22} , \dots,a_{2n} > +\dots+ k_m <a_{m1} ,a_{m2} , \dots,a_{mn} > = 0$

2. The m rows will be linearly independent if the sum in step #1 equals zero

   iff the m number of k's are all equal to zero; or,   $k_1 = k_2 = \dots = k_m = 0,$

3. if there is at least one $k \neq 0$ such that the sum = 0, then try step #1 with only

   the first $m - 1$ rows. We do this decreasing the number of rows until we get independency.

4. It is clear that  $r \leq m$.

5. We also could define rank in terms of the echelon form; so we will study this material in terms of the echelon form.

IV. Given matrix $A = (a_{ij})_{m \times n}$ with rank $r = m$ ; Find this rank from the Echelon Form:

1. echelon form $A' = E\ A = $

$$\begin{pmatrix} 1 & a'_{1\,2} & a'_{1\,3} & \cdots & a'_{1\,n-1} & a'_{1\,n} \\ 0 & 1 & a'_{2\,3} & \cdots & a'_{2\,n-1} & a'_{2\,n} \\ 0 & 0 & 1 & \cdots & a'_{3\,n-1} & a'_{3\,n} \\ \multicolumn{6}{c}{\dotfill} \\ \multicolumn{6}{c}{\dotfill} \\ 0 & 0 & 0 & & 1 & a'_{m-1\,n} \\ 0 & 0 & 0 & 0 & & 1 \end{pmatrix}$$ this 1 is $a'_{m\,n}$

2. testing: $k_1 <1, a'_{1\,2}, a'_{1\,3}, \ldots, a'_{1\,n-1}, a'_{1\,n}> + k_2 <0, 1, a'_{2\,3}, \ldots, a'_{2\,n-1}, a'_{2\,n}> +$

$k_3 <0, 0, 1, \ldots, a'_{3\,n-1}, a'_{3\,n}> + \ldots + k_{m-1} <0, 0, 0, \ldots, 1, a'_{m-1\,n}> +$

$k_m <0, 0, 0, 0, \ldots, 1> \qquad = \qquad <0, 0, 0, \ldots, 0>$ .

3. $<k_1, k_1\ a'_{1\,2}, \ldots, k_1\ a'_{1\,n}> + <0, k_2, k_2\ a'_{2\,3}, \ldots, k_2\ a'_{2\,n}> + \ldots +$

$<0, 0, 0, 0, \ldots, k_m> \qquad = \qquad <0, 0, 0, \ldots, 0>$ .

4. $< k_1 + 0 + 0 + \ldots + 0, k_1\ a'_{1\,2} + k_2 + 0 + \ldots + 0, \ldots, k_1\ a'_{1\,n} + k_2\ a'_{2\,n} + \ldots + k_m > = 0$

5. so:  $k_1 \qquad + 0 \qquad + \ldots + 0 \quad = 0$      or    $k_1 = 0$

       $k_1\ a'_{1\,2} + k_2 \qquad + \ldots + 0 \quad = 0$      &    $k_2 = 0$

       $\dotfill$        $\dotfill$

       $\dotfill$        $\dotfill$

       $k_1\ a'_{1\,n} + k_2\ a'_{2\,n} + \ldots + k_m \quad = 0$      &    $k_m = 0$

6. We see that this echelon form of the matrix A has m linearly independent rows; so the Rank of the echelon form = m.  Thus, the rank of the echelon form is r the same as the row rank of A,

7. It is seen that in order to get independency in the echelon form each row vector has to have "1" as one of its coordinates and be preceded only by 0's . Each "1" in a row has to be to the right of the "1" in the preceding row. This makes k  = 0; then the following k's will also become 0's and thus  force the independency,

8. The number of non-zero rows in the echelon form will be the same as the number of independent rows,

9. If r<m , then the extra row(s) in A' are all zero; so have no effect on the rank .

    If n>m , then the extra column(s) will have no effect on the work to find rank.

10. Note: in the echelon form it would not be possible to have more non-zero rows than there are rows.  We get r < m when the arithmetic gives us zeros that were not part of our plan of getting 1's & 0's in the right places for the echelon form.

11.. We have proved the Theorem: The Row Rank r of a matrix $A = (a_{ij})_{m \times n}$     is

    the Maximum Number of Non-zero Rows in the Echelon Form, and  $r \leq m$ .

 V.

   1.  We define the column rank as the maximum number of linearly independent column matrices.

   2. We then could use the normal way to test for linearly independent columns. We use the idea in step IV. to get the column rank only use columns instead of rows.

 VI.

   1. It has been discovered that the column rank is the same as the row rank; so for convenience we take the row rank as the rank of the matrix. We need to prove that they are the same, for this equality is not obvious. There is more than one way to prove this. Each has its advantages and its disadvantages. We will choose the proof in section VII.

   2. The method dependents on changing the form of the given matrix, $A = (a_{ij})_{m \times n}$      ,to a different one to allow us to see some relations that are not obvious in the original form. Unfortunately this forces the proof to become a little leggy. This happens quite often in Linear Algebra, but if the student will use pencil and paper to rewrite the proof in his own words, the conclusions should become clear.


VII. Theorem: Given: matrix $A = (a_{ij})_{m \times n}$     with rank A = $r \leq m$,

            Then : the Column Rank of A = the Row Rank, r , of A.

   Proof:

      1. If r<m then there are rows in A that are not part of the independent set of rows. In our proof and in most of our work we place these extra rows below the rth row .

2. 
$$A = \begin{pmatrix} a_{11} & a_{12} & \dots & a_{1n} \\ a_{21} & a_{22} & \dots & a_{2n} \\ \dots\dots\dots\dots\dots\dots\dots\dots \\ \dots\dots\dots\dots\dots,\dots\dots \\ a_{r1} & a_{r2} & \dots & a_{rn} \\ a_{r+1\,1} & a_{r+1\,2} & \dots & a_{r+1\,n} \\ \dots\dots\dots\dots\dots\dots\dots\dots \\ \dots\dots\dots\dots\dots\dots\dots\dots \\ a_{m1} & a_{m2} & \dots & a_{mn} \end{pmatrix} \qquad Or \quad A = \begin{pmatrix} a_1 \\ a_2 \\ . \\ . \\ a_r \\ a_{r+1} \\ . \\ . \\ a_m \end{pmatrix}$$

if r = m, then $a_{rj}$ is in the last row; $a_{r+1}$ , $a_{r+2}$ , ... do not exist.

3. The maximum number of these rows that are linearly independent is r. That means that a linear combination of r of the linear independent rows can be used to form every row of A. Let the set of linear independent rows be a matrix that we will label B; so B Spans the matrix A and is a Basis of A.

4.a.  Let  B = ( $b_i$  : i = 1,2, ... ,r )

where each  $b_i$  = < $b_{i\,1}$ , $b_{i\,2}$  , ..., $b_{i\,n}$ > : i = 1,2, ...,r

note:  $b_i$ = $a_i$ :  i = 1,2,...,r  and  $b_{i\,j}$ = $a_{i\,j}$ :  i =1,...,r  &  j = 1,2,...,n

4.b.

$$so: A = \begin{pmatrix} b_1 \\ b_2 \\ . \\ . \\ b_r \\ a_{r+1} \\ . \\ . \\ a_m \end{pmatrix} = \begin{pmatrix} b_{1\,1} & b_{1\,2} & \dots & b_{1\,n} \\ b_{2\,1} & b_{2\,2} & \dots & b_{2\,n} \\ \dots\dots\dots\dots\dots\dots\dots \\ \dots\dots\dots\dots\dots\dots\dots \\ b_{r\,1} & b_{r\,2} & \dots & b_{r\,n} \\ a_{r+1\,1} & a_{r+1\,2} & \dots & a_{r+1\,n} \\ \dots\dots\dots\dots\dots\dots\dots \\ \dots\dots\dots\dots\dots\dots\dots \\ a_{m\,1} & a_{m\,2} & \dots & a_{m\,n} \end{pmatrix}$$

5. Because B is a Basis  of A , linear combinations of the $b_i$ 's  can be used to write each row  $a_i$  :  i = 1,2, ... ,m  of  A.

or: $a_1 = k_{11}\ b_1 + k_{12}\ b_2 + \ldots + k_{1n}\ b_r$        where the k's are constants;

$a_2 = k_{21}\ b_1 + k_{22}\ b_2 + \ldots + k_{2n}\ b_r$        we note that while not important,

.........................................        for $a_1$ we see that $k_{11} = 1$

.....................................        while all the other k's $= 0$ ;etc.

$a_r = k_{r1}\ b_1 + k_{r2}\ b_2 + \ldots + k_{rn}\ b_r$        note this will end here at $a_r$ if $r = m$,

.....................................

.....................................

$a_{1n} = k_{m1}\ b_1 + k_{m2}\ b_2 + \ldots + k_{mn}\ b_r$

and in components:

$a_1 = k_{11}\ \langle b_{11}, b_{12}, \ldots, b_{1n}\rangle + k_{12}\ \langle b_{21}, b_{22}, \ldots, b_{2n}\rangle + \ldots + k_{1n}\ \langle b_{r1}, b_{r2}, \ldots, b_{rn}\rangle$

$a_2 = k_{21}\ \langle b_{11}, b_{12}, \ldots, b_{1n}\rangle + k_{22}\ \langle b_{21}, b_{22}, \ldots, b_{2n}\rangle + \ldots + k_{2n}\ \langle b_{r1}, b_{r2}, \ldots, b_{rn}\rangle$

...........................................................................

...........................................................................

$a_m = k_{m1}\ \langle b_{11}, b_{12}, \ldots, b_{1n}\rangle + k_{m2}\ \langle b_{21}, b_{22}, \ldots, b_{2n}\rangle + \ldots + k_{mn}\ \langle b_{r1}, b_{r2}, \ldots, b_{rn}\rangle$

6. We are starting to see what the "leggy" statement is all about. The k's are different constants that vary with each $b_i$ and with each b's components, $b_{ij}$. The double subscript has no special mathematical significance here, and is used to show that in general they are all different numbers. The double subscript on a $b_{ij}$ is used to clearly show which member, $b_i$, of the independent set it belongs to and which component of $b_i$ it represents.

7.  $a_1 = \langle k_{11}\,b_{11}, k_{11}b_{12}, \ldots, k_{11}b_{1n}\rangle + \langle k_{12}\,b_{21}, k_{12}b_{22}, \ldots, k_{12}b_{2n}\rangle + \ldots$

$+ \langle k_{1n}\,b_{r1}, k_{1n}\,b_{r2}, \ldots, k_{1n}\,b_{rn}\rangle$

$a_2 = \langle k_{21}\,b_{11}, k_{21}b_{12}, \ldots, k_{21}b_{1n}\rangle + \langle k_{22}\,b_{21}, k_{22}b_{22}, \ldots, k_{22}b_{2n}\rangle + \ldots$

$+ \langle k_{2n}\,b_{r1}, k_{2n}\,b_{r2}, \ldots, k_{2n}\,b_{rn}\rangle$

...........................................................................

...........................................................................

$a_m = \langle k_{m1}\,b_{11}, k_{m1}\,b_{12}, \ldots, k_{m1}b_{1n}\rangle + \langle k_{m2}\,b_{21}, k_{m2}b_{22}, \ldots, k_{m2}b_{2n}\rangle + \ldots$

$+ \langle k_{mn}\,b_{r1}, k_{mn}\,b_{r2}, \ldots, k_{mn}\,b_{rn}\rangle$

8. adding each row matrix to get a different form for each $a_i$ :

$$a_1 = \ < \ [k_{11} b_{11} + k_{12} b_{21} \ , ..., \ k_{1n} b_{r1} \ ] \ , \ [k_{11} b_{12} + k_{12} b_{22} + ... + k_{1n} b_{r2} \ ] \ , ...,$$

$$[k_{11} b_{1n} + k_{12} b_{2n} + ... + k_{1n} b_{rn} \ ] >$$

$$a_2 = \ < \ [k_{21} b_{11} + k_{22} b_{21} + ... + k_{2n} b_{r1} \ ] \ , \ [k_{21} b_{12} + k_{22} b_{22} + ... + k_{2n} b_{r2} \ ] \ , ...,$$

$$[k_{21} b_{1n} + k_{22} b_{2n} + ... + k_{2n} b_{rn} \ ] >$$

$$.............................................................................$$

$$.............................................................................$$

$$a_m = \ < \ [k_{m1} b_{11} + k_{m2} b_{21} + ... + k_{mn} b_{r1} \ ] \ , \ [k_{m1} b_{12} + k_{m2} b_{22} + ... + k_{mn} b_{r2} \ ] \ , ...,$$

$$[k_{m1} b_{1n} + k_{m2} b_{2n} + ... + k_{mn} b_{rn} \ ] > \ \ .$$

9. Write the first two columns of $a_{11}$  as the sum of n columns:

|  | 1$^{st}$ term | 2$^{nd}$ term | nth term |
|---|---|---|---|

first column:

$$\begin{pmatrix} k_{11} b_{11} \\ k_{21} b_{11} \\ . \\ . \\ k_{m1} b_{11} \end{pmatrix} + \begin{pmatrix} k_{12} b_{21} \\ k_{22} b_{21} \\ . \\ . \\ k_{m2} \ b_{21} \end{pmatrix} + ... + \begin{pmatrix} k_{1n} b_{r1} \\ k_{2n} b_{r1} \\ . \\ . \\ k_{mn} b_{r1} \end{pmatrix}$$

second column:

$$\begin{pmatrix} k_{11} b_{12} \\ k_{21} b_{12} \\ . \\ . \\ k_{m1} b_{12} \end{pmatrix} + \begin{pmatrix} k_{12} b_{22} \\ k_{22} b_{22} \\ . \\ . \\ k_{m2} b_{22} \end{pmatrix} + ... + \begin{pmatrix} k_{1n} b_{r2} \\ k_{2n} b_{r2} \\ . \\ . \\ k_{mn} b_{r2} \end{pmatrix}$$

10. It is clear from step 9 that the general jth column of $a_{11}$  is the  sum:

jth column:

$$\begin{pmatrix} k_{11} b_{1j} \\ k_{21} b_{1j} \\ . \\ . \\ k_{m1} b_{1j} \end{pmatrix} + \begin{pmatrix} k_{12} b_{2j} \\ k_{22} b_{2j} \\ . \\ . \\ k_{m2} \ b_{2j} \end{pmatrix} + ... + \begin{pmatrix} k_{1n} b_{rj} \\ k_{2n} b_{rj} \\ . \\ . \\ k_{mn} b_{rj} \end{pmatrix}$$

for our purpose this can be written in the better from of:

jth column:

$$\begin{pmatrix} k_{1\,1} \\ k_{2\,1} \\ . \\ . \\ k_{m\,1} \end{pmatrix} b_{1\,j} \;+\; \begin{pmatrix} k_{1\,2} \\ k_{2\,2} \\ . \\ . \\ k_{m\,2} \end{pmatrix} b_{2\,j} \;+\; \dots \;+\; \begin{pmatrix} k_{1\,n} \\ k_{2\,n} \\ . \\ . \\ k_{m\,n} \end{pmatrix} b_{r\,j}$$

It is obvious that except for the constants $k_{i\,j}$ this represents every column of A [see 4b]

11. Every column of A is in the form of a linear combination of r number of the  b's  --- the set

of linearly independent rows. Thus, the Column Rank can not be greater than r;

the Column Rank of A [ $r_C$ (A)] <  the Row Rank of A [ $r_R$ (A)] ,

or using  symbolic notation:                    $r_C$ (A) $\leq r_R$ (A) = r

12.a. Note: While we proved this with matrix A ,

it is true of any matrix --- even a transpose, as        $r_C$ $(A^T)$ $\leq$ $r_R$ $(A^T)$ ,

b. Now the Rows of A are the same as

the Columns of A transpose; so        $r_C$ $(A^T)$ = $r_R$ (A).

c.The Columns of A are the same as

the Rows of A transpose; so        $r_R$ $(A^T)$ = $r_C$ (A).

13. Substituting                    $r_R$ (A ) $\leq$ $r_C$ (A ) $\leq$ $r_R$ (A)

14. In words:      Row Rank of A $\leq$ Column Rank of A $\leq$ Row Rank of A    ;

this can only be true where :     the Column Rank of a matrix equals the Row Rank.

15. With this  equality  we define the common column and row rank of the matrix as the

Rank of the Matrix.


VIII.

1. Generally speaking it is easier to find rank from the echelon form rather than

finding a linear independent set of rows as in section 9.7 - III. 1-4. Later we will see

still another way to do this.

2. Note:  Rank of the Zero Matrix is 0; and the Rank of A = 0  iff A = 0.

IX. Example: Find the Rank of

$$A = \begin{bmatrix} 3 & 1 & -8 & -2 \\ -1 & 0 & 4 & 3 \\ 5 & 2 & -12 & -1 \end{bmatrix}$$

1.

$$\begin{bmatrix} 3 & 1 & -8 & -2 \\ -1 & 0 & 4 & 3 \\ 5 & 2 & -12 & -1 \end{bmatrix} \updownarrow \longrightarrow \begin{bmatrix} -1 & 0 & 4 & 3 \\ 3 & 1 & -8 & -2 \\ 5 & 2 & -12 & -1 \end{bmatrix} \quad m[-1]$$

$$\begin{bmatrix} 1 & 0 & -4 & -3 \\ 3 & 1 & -8 & -2 \\ 5 & 2 & -12 & -1 \end{bmatrix} \begin{array}{l} m[-3] \;/\; m[-5] \\ + \\ + \end{array} \longrightarrow \begin{bmatrix} 1 & 0 & -4 & -3 \\ 0 & 1 & 4 & 7 \\ 0 & 2 & 8 & 14 \end{bmatrix} \quad m[-2] \; +$$

$$\begin{bmatrix} 1 & 0 & -4 & -3 \\ 0 & 1 & 4 & 7 \\ 0 & 0 & 0 & 0 \end{bmatrix}$$

2. The number of non-zero rows is two; so the Rank is 2.

3. as a check of this test for independency : set   $c\langle 1,0,-4,-3\rangle + k\langle 0,1,4,7\rangle = 0$

   now:  $\langle\, c\,,\, k\,,\, -4c+4k\,,\, -3c+7k\,\rangle = \langle 0,0,0,0\rangle$  iff  $c = 0$  &  $k = 0$ ; so Independent

4. Exercise: Using the A of IX. To prove that any two rows are Linearly Independent.


8.8. The INVERSE of a SQUARE MATRIX

   I. Introduction  [ the Inverse can be used in the Solution of Systems ].

   1. In the Real Number System the concept of a Product lead us into the concept of
Division. Given the equation involving the product, $r\,t = 1$, then t could obtained by a
division symbolized by a fraction, as : $t = 1/r$ if $r = 0$. We also called the $1/r$ the

Reciprocal of t. In a more rigorous treatment of the mathematics we introduced the

concept of the Inverse to handle this concept.

2. If $t = 1/r$ & $r \neq 0$, we invented a number called the "Inverse of r" and used the

written form $r^{-1}$ such that $r \; r^{-1} = r^{-1} \; r = 1$. Note here the superscript $-1$ is not

an exponent. The $r^{-1}$ is a new symbol; it does not normally get confused with exponents.

The commutative property of this definition is important to the concept so is a

necessary part of the definition.

3. With this concept in the real number system we turned a division $r/t$ into

a product $r \; (1/t)$ and then into a better product $r \; t^{-1}$ .

II.

1. In Matrix Theory we find that we have serious problems if we try to deal with this

concept by means of a real number system division operation; so we take the inverse

route to handle this concept.

2. The question arises whether given a matrix $A \neq 0$ is there a matrix $B \neq 0$ such that

$A \; B = B \; A = I_n$ where $I_n$ is the unit matrix :

$$\begin{bmatrix} 1 & 0 \\ 0 & 1 \end{bmatrix} \text{ or } \begin{bmatrix} 1 & 0 & 0 \\ 0 & 1 & 0 \\ 0 & 0 & 1 \end{bmatrix} \text{ or } \begin{bmatrix} 1 & 0 & 0 & 0 \\ 0 & 1 & 0 & 0 \\ 0 & 0 & 1 & 0 \\ 0 & 0 & 0 & 1 \end{bmatrix} \text{ or } \ldots,$$

and thus B be the Inverse of A ?

3. We saw in 8.6 how we could change by elementary operations a matrix all the way to I.

In 8.5 we saw how we could represent a series of such operations as the product of

matrices; so part of the answer is yes. We need, however, to set up the concept of the

inverse more formally.

4. Because it is not possible for a Matrix Product to be commutative unless the matrices

are Square, we restrict this study to Square Matrices.

III. Definition: The Inverse of a Square Matrix $A_{n \times n} = A_n = [a_{ij}]$ : i & j $= 1, \ast\ast\ast, n$ is

symbolized by $A^{-1}$ and satisfies the equations: $A \; A^{-1} = A^{-1} \; A = I_n$ where $A \neq 0$.

IV. Examples:

1. Let $A = \begin{bmatrix} 1 & 4 \\ 2 & 9 \end{bmatrix}$ and $B = \begin{bmatrix} 9 & -4 \\ -2 & 1 \end{bmatrix}$

2. then: $AB = \begin{bmatrix} 1 & 4 \\ 2 & 9 \end{bmatrix} \begin{bmatrix} 9 & -4 \\ -2 & 1 \end{bmatrix} = \begin{bmatrix} [9-8] & [-4+4] \\ [18-18] & [-8+9] \end{bmatrix} = \begin{bmatrix} 1 & 0 \\ 0 & 1 \end{bmatrix} = I_2$

  so $AB = I$   implies   $B = A^{-1}$

3. note $BA = \begin{bmatrix} 9 & -4 \\ -2 & 1 \end{bmatrix} \begin{bmatrix} 1 & 4 \\ 2 & 9 \end{bmatrix} = \begin{bmatrix} [9-8] & [36-36] \\ [-2+2] & [-8+9] \end{bmatrix} = \begin{bmatrix} 1 & 0 \\ 0 & 1 \end{bmatrix} = I_2$

  so $BA = I$   implies $A = B^{-1}$

V. Because of the commutative property of the product of these two matrices equaling the identity matrix we see that either matrix is the inverse of the other. From $A^{-1} A = I$ we see that the second factor A is the inverse of the first. So, $[A^{-1}]^{-1} = A$.

VI. Examples.

  A.

   1. Examine the zero matrix; or given $A \begin{bmatrix} 0 & 0 \\ 0 & 0 \end{bmatrix}$,

   2. Assume $\begin{bmatrix} x & y \\ u & v \end{bmatrix}$ is the inverse of A , then

   $\begin{bmatrix} 0 & 0 \\ 0 & 0 \end{bmatrix}\begin{bmatrix} x & y \\ u & v \end{bmatrix} = \begin{bmatrix} 0 & 0 \\ 0 & 0 \end{bmatrix}$ which is not equal to I  or  $\begin{bmatrix} 1 & 0 \\ 0 & 1 \end{bmatrix}$

   3. so A = 0 has no Inverse.

  B.  Find the inverse of $A = \begin{bmatrix} 2 & 1 \\ 2 & 1 \end{bmatrix}$

   1. $\begin{bmatrix} 2 & 1 \\ 2 & 1 \end{bmatrix}\begin{bmatrix} x & y \\ z & w \end{bmatrix} = \begin{bmatrix} 1 & 0 \\ 0 & 1 \end{bmatrix}$ or $\begin{bmatrix} [2x+z] & [2y+w] \\ [2x+z] & [2y+z] \end{bmatrix} = \begin{bmatrix} 1 & 0 \\ 0 & 1 \end{bmatrix}$

   2. and for this to be equal to $I_2$ $\begin{cases} 2x+z = 1 \\ 2x+z = 0 \end{cases}$ & $\begin{cases} 2y+w = 0 \\ 2y+w = 1 \end{cases}$ [from row 1] [from row 2]

3. Clearly the systems are inconsistent; so no solution and

      this A does not have an inverse.

C. Find the inverse of A = $\begin{bmatrix} 1 & 1 \\ 1 & -1 \end{bmatrix}$

1. $\begin{bmatrix} 1 & 1 \\ 1 & -1 \end{bmatrix} \begin{bmatrix} x & y \\ z & w \end{bmatrix} = \begin{bmatrix} 1 & 0 \\ 0 & 1 \end{bmatrix}$

2. so $\left\{ \begin{array}{l} x + z = 1 \\ x - z = 0 \end{array} \right\}$ & $\left\{ \begin{array}{l} y + w = 0 \\ y - w = 1 \end{array} \right\}$

3. adding $2x = 1$ & $2y = 1$

   or $x = \frac{1}{2}$ & $z = \frac{1}{2}$        $y = \frac{1}{2}$ & $w = -\frac{1}{2}$

4. thus A = $\begin{bmatrix} \frac{1}{2} & \frac{1}{2} \\ \frac{1}{2} & -\frac{1}{2} \end{bmatrix}$

VII. Existence of the Inverse of a Square Matrix.

1. To study this existence possibility in more detail we will consider the example in section 8.6. IV. where we saw that A = $\begin{bmatrix} 1 & 0 & -4 \\ 3 & 1 & -8 \\ 5 & 2 & 10 \end{bmatrix}$ ~ $\begin{bmatrix} 1 & 0 & 0 \\ 0 & 1 & 0 \\ 0 & 0 & 1 \end{bmatrix}$ = $I_3$

2. The rank of A is 3, and that is also the rank of $I_3$. We note that the number of rows [ and of the columns ] is 3.

3. It is clear that if the rank of a square matrix is the same as the number of the rows [ or columns ] , we can use a series of elementary operations to arrive at the equivalent reduced echelon form which is $I_n$    .

Or, $E_n$  [... { $E_2$  ($E_1$  A )} ]  =  E A  = $I_n$   ; so this E is the inverse $A^{-1}$   of A.

4. Thus, we have a constructive proof of the theorem that if the rank of a square matrix is the same as the number of rows [or columns] , then the Inverse of that matrix does exist.

8.9 Definitions.

I. If a Square Matrix has no Inverse it is a Singular Matrix    -- as in being unusual, rare, odd, or fitting a particular, special case.

II. If a Square Matrix has an Inverse, it is a Non-Singular Matrix. Caution: one's memory can play tricks here for a positive thing  [has an Inverse] has a negative name [non-singular].

8.10  Theorems and an Example.

   I. If A has an Inverse  $A^{-1}$ , then  it is Unique.

    Proof:

      1. given A  $A^{-1}$  = $A^{-1}$ A = I

      2. consider that there is another matrix B such that A B = B A = I

      3. then B = B I = B [ A $A^{-1}$ ] = [B A] $A^{-1}$  = I $A^{-1}$  = $A^{-1}$

      4 . thus there is at most only one matrix, $A^{-1}$  , that satisfies step 1

      5. therefore $A^{-1}$  is the Inverse of A, and $A^{-1}$  is Unique.

II.  Given: A,S, & T as nth order square matrixes

   Then :   A S = A T    implies S = T

   Proof:

     1.        A S = A T

     2. $A^{-1}$ [ A S ] = $A^{-1}$ [ A T ]

        [$A^{-1}$ A ] S  = [$A^{-1}$ A ] T

           I S  =  I T

     3.        S  =   T


III. Example. Find $A^{-1}$    given A = $\begin{bmatrix} 2 & 1 \\ -1 & 1 \end{bmatrix}$

   1.Assume $A^{-1}$   =  $\begin{bmatrix} u & v \\ w & x \end{bmatrix}$

   2. then $\begin{bmatrix} 2 & 1 \\ -1 & 1 \end{bmatrix}$ $\begin{bmatrix} u & v \\ w & x \end{bmatrix}$ = $\begin{bmatrix} 1 & 0 \\ 0 & 1 \end{bmatrix}$

   3. $\begin{bmatrix} [2u+w] & [2v+x] \\ [-u+w] & [-v+x] \end{bmatrix}$ = $\begin{bmatrix} 1 & 0 \\ 0 & 1 \end{bmatrix}$

   4. $\begin{cases} 2u+w = 1 \\ -u+w = 0 \end{cases}$    &    $\begin{cases} 2v+x = 0 \\ -v+x = 1 \end{cases}$

   5.    3u  = 1    &     3v  = -1           { substracting }

         u  = 1/3 &  v  = -1/3

6. into 2$^{nd}$ equations:   w = 1/3   &   x = 1 - 1/3 = 2/3

7. A$^{-1}$   =   $\begin{bmatrix} 1/3 & -1/3 \\ 1/3 & 2/3 \end{bmatrix}$


8.11 To Calculate Inverses.

   I.

      1. As the order of the matrix gets larger, it would seem that the method to find inverses used above would get considerably harder; so we will now hunt for an improvement to the method.

      2. look at the some of the shorthand equations  for indicating the system in 9.12:

         a. now     A  X = D        but D can also be written as          D = I D

         b. now     I  X = X        but from the first of a.  X is        X =     A$^{-1}$  D

      3. it is seems that A is connected to I in the same way that I is

            connected to A$^{-1}$    .

      4. so if A is changed by a set of matrix elementary row operations to the form of I , then that set should change the form of I to A$^{-1}$       --- note the arrangement in the following example that can be used to keep track of the changes.


II. Example. Find A$^{-1}$      if A  =  $\begin{bmatrix} 3 & 4 \\ 2 & 3 \end{bmatrix}$

1. take     A            I

A | I =  $\begin{bmatrix} 3 & 4 & | & 1 & 0 \\ 2 & 3 & | & 0 & 1 \end{bmatrix}$   m[-2/3]      →      $\begin{bmatrix} 3 & 4 & | & 1 & 0 \\ 0 & 1/3 & | & -2/3 & 1 \end{bmatrix}$   m[3]

$\begin{bmatrix} 3 & 4 & | & 1 & 0 \\ 0 & 1 & | & -2 & 3 \end{bmatrix}$   m[-4]   →   $\begin{bmatrix} 3 & 0 & | & 9 & -12 \\ 0 & 1 & | & -2 & 3 \end{bmatrix}$   m[1/3]

$\begin{bmatrix} 1 & 0 & | & 3 & -4 \\ 0 & 1 & | & -2 & 3 \end{bmatrix}$

        I            A$^{-1}$

3. Check:    $A^{-1}\ A\ =\ \begin{bmatrix} 3 & -4 \\ -2 & 3 \end{bmatrix} \begin{bmatrix} 3 & 4 \\ 2 & 3 \end{bmatrix}\ =\ \begin{bmatrix} [9-8] & [12-12] \\ [-6+6] & [-8+9] \end{bmatrix}\ =\ \begin{bmatrix} 1 & 0 \\ 0 & 1 \end{bmatrix}\ =\ I_n.$

III. note: if   $A\ =\ \begin{bmatrix} 1 & 1 \\ -3 & -3 \end{bmatrix}$   &   $B\ =\ \begin{bmatrix} 0 & -1 \\ -1 & 1 \end{bmatrix},$

then   $A\ B\ =\ \begin{bmatrix} 1 & 1 \\ -3 & -3 \end{bmatrix} \begin{bmatrix} 1 & -1 \\ -1 & 1 \end{bmatrix}\ =\ \begin{bmatrix} 0 & 0 \\ 0 & 0 \end{bmatrix}\ =\ O$

so   $A\ B\ =\ O$   does not imply that either $A = O$   or   $B = O$ .

8.12. Theorem. Prove    $[\ A\ B\ ]^{-1}\ =\ B^{-1}\ A^{-1}$        {note the reverse order}

   1. AB is a matrix; so    $[AB]\ [AB]^{-1}\ \ =\ I$

   2. $A^{-1}\ \{[AB]\ [AB]^{-1}\ \}\ \ \ =\ A^{-1}\ \ \ I$

      $\{[A^{-1}\ A]\}\ [B]\ [AB]^{-1}\ \ \ =\ A^{-1}$

      $[I\ B]\ [AB]^{-1}\ \ \ =\ A^{-1}$

      $B\ [AB]^{-1}\ \ \ =\ A^{-1}$

   3. multiply by $B^{-1}$    :    $B^{-1}\ \{\ [B]\ [AB]^{-1}\ \}\ \ \ =\ B^{-1}\ A^{-1}$

            $[B^{-1}\ B]\ [AB]^{-1}\ \ \ =\ B^{-1}\ A^{-1}$

            $I\ \ [AB]^{-1}\ \ \ =\ B^{-1}\ A^{-1}$

            $[AB]^{-1}\ \ \ =\ B^{-1}\ A^{-1}$            QED.

   4. This can be extended to:   $[ABC\ ....P]^{-1}\ =\ P^{-1}\ ....\ C^{-1}\ B^{-1}\ A^{-1}$   .

8.13. A First Degree System where the matrix A is Square.

    I. Given a nth order [nxn] system of first degree equations; find the solution.

        1. given:

$$\begin{cases} a_{11}\ \ x_1\ + a_{12}\ \ x_2\ + \ldots + a_{1n}\ \ x_n\ = c_1 \\ a_{21}\ \ x_1\ + a_{22}\ \ x_2\ + \ldots + a_{2n}\ \ x_n\ = c_2 \\ \ldots\ldots\ldots\ldots\ldots\ldots\ldots\ldots\ldots\ldots\ldots\ldots \\ \ldots\ldots\ldots\ldots\ldots\ldots\ldots\ldots\ldots\ldots\ldots\ldots \\ a_{n1}\ \ x_1\ + a_{n2}\ \ x_2\ + \ldots + a_{nn}\ \ x_n\ = c_n \end{cases}$$

        2. this can be written as a matrix equation:

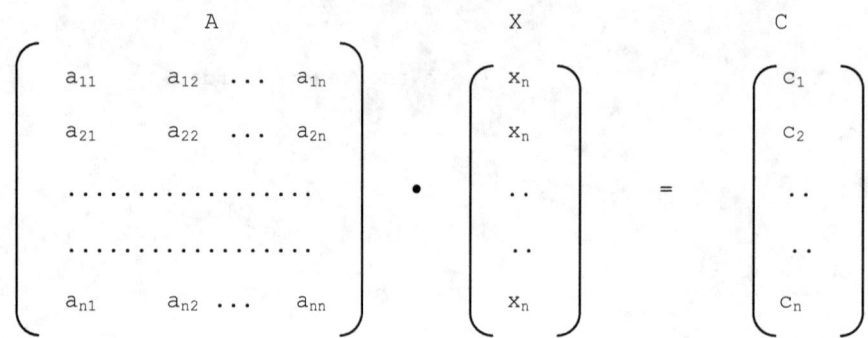

        3. or $\ A\ X\ = C$

           where $\ A = (\ a_{ij}\ )_{n \times n}\ ;\ X = (\ x_{ij}\ )_{n \times 1}\ ;\ C = (\ c_{ij}\ )_{n \times 1}$

        4. Assume A is Non-Singular; thus A has an Inverse $A$ ,

        5. now $A^{-1}\ \ C\ = A^{-1}\ [\ A\ X\ ] = [A^{-1}\ A]\ X = I_n\ X\ =\ X$

        6.   The Solution Set:  $X\ = A^{-1}\ \ C$

II.

  A. We have proved the Theorem:  Given the system of first degree equations  $A\ X = C$

                  where $\ A = (\ a_{ij}\ )_{nxn}\ \ \ \&\ \ \ C = (\ c_{ij}\ )_{nx1}$

              Then the Solution $\ X\ =\ (\ x_{ij}\ )_{n \times 1}$

                is such that $\ X\ = A^{-1}\ \ C.$

  B. Is the solution unique?

    1. assume that Y is also a solution,

    2. then $\ A\ Y = C$

    3. or $A^{-1}\ [\ A\ Y\ ] = A^{-1}\ C\ \ \ \ $ thus $[\ A^{-1}\ A\ ]\ Y\ = A^{-1}\ C\ \ \ \ $ and $I_n\ Y\ = A^{-1}\ \ C$

       therefore $\ \ \ Y\ = A^{-1}\ \ C\ \ \ \ \ \ $ so $\ \ \ \ \ A^{-1}\ \ C$ is still the solution.

8.14 Back to solving Systems of First Degree Equations.

I. In order to see how to apply the materials in the previous sections to solve systems we first need to look at the method used to do this in the previous elementary algebra. In 8.15 we will use the 8.14 given $3^{rd}$ order system to do this, and then we will apply two matrix methods to solve the same system. We then will generalize to any order.

II. Find the solution set given the following $3^{rd}$ order system:

$$x_1 + 2x_2 - x_3 = -2 \qquad \alpha$$
$$2x_1 + x_2 + x_3 = 1 \qquad \beta$$
$$3x_1 - x_2 + 2x_3 = 4 \qquad \gamma$$

8.15 Solutions.

I. One form of the elementary algebra method of solution.

1. change this $3^{rd}$ order system into a $2^{nd}$ order system that is equivalent/equal to the given ; this means it has the same solution set as the given.

$$\alpha : m[-2] : \quad -2x_1 - 4x_2 + 2x_3 = 4$$
$$\beta \qquad : \quad 2x_1 + x_2 + x_3 = 1$$
$$\text{add} \qquad : \qquad -3x_2 + 3x_3 = 5 \qquad \delta$$

$$\alpha : m[-3] : \quad -3x_1 - 6x_2 + 3x_3 = 6$$
$$\gamma \qquad : \quad 3x_1 - x_2 + 2x_3 = 4$$
$$\text{add} \qquad : \qquad -7x_2 + 5x_3 = 10 \qquad \phi$$

so the $2^{nd}$ order equivalent :

$$-3x_2 + 3x_3 = 5 \qquad \delta$$
$$-7x_2 + 5x_3 = 10 \qquad \phi$$

2. change this 2nd order system into an equivalent/equal [same solution set]1st order :

$$\delta : m[-7/3]: \qquad 7x_2 - 7x_3 = -35/3$$
$$\phi : \qquad -7x_2 + 5x_3 = 10$$
$$\text{add} \qquad : \qquad -2x_3 = -5/3$$
$$x_3 = 5/6$$

3. substituting into $\delta$ to get $x_2 = -5/6$

4. substituting both into $\alpha$ to get $x_1 = 1/2$

5. thus the solution set is : ( ½ , -5/6 , 5/6 )

II.

1. What we did basically in the solution of this system  was to use elementary algebraic methods to first eliminate $x_1$  from our system without changing the solution set and then eliminating $x_2$   to arrive at a single equation in only $x_3$   . The rest was simple and obvious.

2. There are a couple of ways to do this ; we chose the way where the $3^{rd}$ order system was reduced to a $2^{nd}$ order which in turn was reduced to a $1^{st}$ order.

3. This is called the Reduction Method of solving systems. Some writers call this the Gauss Elimination method.

4. What we want to do now is to show how to do this  with matrices. We will start with the method that uses the echelon form of a matrix.


8.16 Echelon method of solving the 8.15 system.

I.

1. We start with the Augmented Matrix form of the system:
$$\begin{bmatrix} 1 & 2 & -1 & -2 \\ 2 & 1 & 1 & 1 \\ 3 & -1 & 2 & 4 \end{bmatrix}$$

2. now convert to Echelon Form:

$$m[-3] \; / \; m[-2] \quad \begin{bmatrix} 1 & 2 & -1 & -2 \\ 2 & 1 & 1 & 1 \\ 3 & -1 & 2 & 4 \end{bmatrix} \rightarrow \begin{bmatrix} 1 & 2 & -1 & -2 \\ 0 & -3 & 3 & 5 \\ 0 & -7 & 5 & 10 \end{bmatrix} \quad m[-1/3] \rightarrow \begin{bmatrix} 1 & 2 & -1 & -2 \\ 0 & 1 & -1 & -5/3 \\ 0 & -7 & 5 & 10 \end{bmatrix}$$

$$\begin{bmatrix} 1 & 2 & -1 & -2 \\ 0 & 1 & -1 & -5/3 \\ 0 & -7 & 5 & 10 \end{bmatrix} \quad m[7] \qquad \longrightarrow \qquad \begin{bmatrix} 1 & 2 & -1 & -2 \\ 0 & 1 & -1 & -5/3 \\ 0 & 0 & -2 & -5/3 \end{bmatrix} \quad m[-1/2]$$

$$\begin{bmatrix} 1 & 2 & -1 & -2 \\ 0 & 1 & -1 & -5/3 \\ 0 & 0 & 1 & 5/6 \end{bmatrix} \qquad ; \text{ this is the echelon form  of the augmented matrix of the system.}$$

3.  Convert it back to an equation system form:

$$\left\{\begin{array}{llll} x_1 & + \ 2x_2 & -x_3 & = \ -\ 2 \\ & x_2 & -x_3 & = \ -5/3 \\ & & x_3 & = \ 5/6 \end{array}\right\} \qquad \begin{array}{l} \alpha \\ \\ \beta \\ \\ \delta \end{array}$$

4. $\delta$ into the $2^{nd}$ equation:   $x_2 = -5/3 + 5/6 = -5/6$; $x_2$  &  $\delta$ into the $1^{st}$ equation:

 $x_1 = -\ 2 + 5/3 + 5/6 = 1/2$   ;the solution set:  ( 1/2 ,   -5/6,   5/6 )

5. This is the  Gauss Elimination Method adapted to Matrices. Some writers say or infer

that the Gauss method is a matrix method, but Gauss died in 1855 and the first use of

the word matrix was 1850 and there was very little theory developed until after, say,

1858. Gauss's enormous genius was in other mathematical/physical fields. Note that this

matrix method eliminates the superfluous: the x's , the +'s , the ='s. So, this matrix

method is superior to the elementary algebra method.

6. It is seen that if a matrix is the augmented matrix form of a system of equations,

then the equivalency of the echelon form  means this form has the same solution set. It

is clear that the elementary row operations on the augmented matrix are the same as the

equation axioms applied to the system of equations in the elementary method of solution.

   II. Solution of the same system by use of the reduced echelon form. Start with last

matrix of step #2 of I.

$$\left[\begin{array}{cccc} 1 & 2 & -1 & -\ 2 \\ 0 & 1 & -1 & -5/3 \\ 0 & 0 & 1 & 5/6 \end{array}\right] \begin{array}{l} + \\ m(-2) \end{array} \Rightarrow \left[\begin{array}{cccc} 1 & 0 & 1 & 4/3 \\ 0 & 1 & -1 & -5/3 \\ 0 & 0 & 1 & 5/6 \end{array}\right] \begin{array}{l} + \\ m(1) \end{array}$$

$$\left[\begin{array}{cccc} 1 & 0 & 1 & 4/3 \\ 0 & 1 & 0 & -5/6 \\ 0 & 0 & 1 & 5/6 \end{array}\right] \begin{array}{l} + \\ m(-1) \end{array} \Rightarrow \left[\begin{array}{cccc} 1 & 0 & 0 & \tfrac{1}{2} \\ 0 & 1 & 0 & -5/6 \\ 0 & 0 & 1 & 5/6 \end{array}\right] \Rightarrow \left\{\begin{array}{lll} x_1 & = & 1/2 \\ x_2 & = & -5/6 \\ x_3 & = & 5/6 \end{array}\right\}$$

Some writers consider this method better because there is no "back-subsitution" ; of

course there is more matrix operations.

8.17 Existence of Solutions for First Degree Systems in terms of the matrix, A.

I. In n-order systems the number of columns, n , in A is the number of unknowns , and the number of rows, m , is less or equal to the number of unknowns; so in all cases the dimension of X = n and is $\geq$ m. Note: m can not be greater than n, for the extra equation(s) have to be duplicate(s) of the other equation(s) so can be ignored. Looking at examples of various systems and their echelon forms we see that the following statements are true.

II.

1. Let A be nxn and the rank r = n. Then the last row in the echelon form always will be non-zero and in the equation form is a simple 1 x = c. No matter the size of n this equation gives the value of this x as one and only one number. Then substituting this x into the next to last equation we get an unique value for $x_{n-1}$ . Continuing upward in this manner we get unique values for each x .

2. Therefore, an unique solution set ( $x_1$ ,$x_2$ ,...,$x_n$ ) exists if r = n where A is nxn. This linear system is Consistent [has a solution] and Independent [no one of the equations can be built as a linear combination of the others].

III.

1. If the matrix A has a rank r < n, as in the case where A is 3x3 & r = 2 the final step would look like this:


echelon form                    system of equations

$$\begin{bmatrix} 1 & a & b & d \\ 0 & 1 & c & e \\ 0 & 0 & 0 & 0 \end{bmatrix} \text{ implies } \left\{ \begin{array}{l} x_1 + a\,x_2 + b\,x_3 = d \\ \phantom{x_1 + a\,} x_2 + c\,x_3 = e \end{array} \right\}$$

whose solution set is found by:


let $x_3$ be any number and then work out $x_2$ & $x_1$ in terms of that number from the other two equations; so there are an infinite number of triplets in the solution set. It is easy to see that this same thing applies for all nxn A's whatever the size, and note also for all mxn A's where m < n . This is the method for finding solutions in the case where r < n;

2. Therefore, there are an infinite number of n-tuples in the solution set if r < n in an nxn matrix A. The solution set will look different depending on which arbitrary value

is chosen for  x  --- this value is called a Parameter.

3.It is clear that the critical thing here is that r<n and not the size of A. So, there

 are an infinite number of n-tuples in the solution set if r < n in any mxn matrix A.

4. This kind of a system whether A is square or not square is Consistent [ has at least

one solution set]  but Dependent [ at least one equation is a linear combination of one

or more of the others ].

III.

 1. Look at the system in 8.13.I.1. If all the c's are 0 then it is  a Homogeneous Linear System.

It is clear that this system has the  trivial solution of $x_1 = x_2 = ... = x_n = 0$; this is the

only solution if r = n. Note that after the echelon form is converted back to equation form the

last equation is of the form $x_n = 0$ which then forces all of the other x's to be 0.

 2. If in a Homogeneous Linear System r < n , then we have a part II. case of an

infinite number of  n-tuples for  the solution set--- trivial case included.


8.18 The Inverse Matrix Method of solving the 8.15 system --- see 8.14. II.

 I.

 1. From 9.15 set up the matrices: A the coefficients, X the unknowns, C the

constants:

$$A = \begin{bmatrix} 1 & 2 & -1 \\ 2 & 1 & 1 \\ 3 & -1 & 2 \end{bmatrix} \qquad X = \begin{bmatrix} x_1 \\ x_2 \\ x_3 \end{bmatrix} \qquad C = \begin{bmatrix} -2 \\ 1 \\ 4 \end{bmatrix}$$


2. Find  $A^{-1}$  :

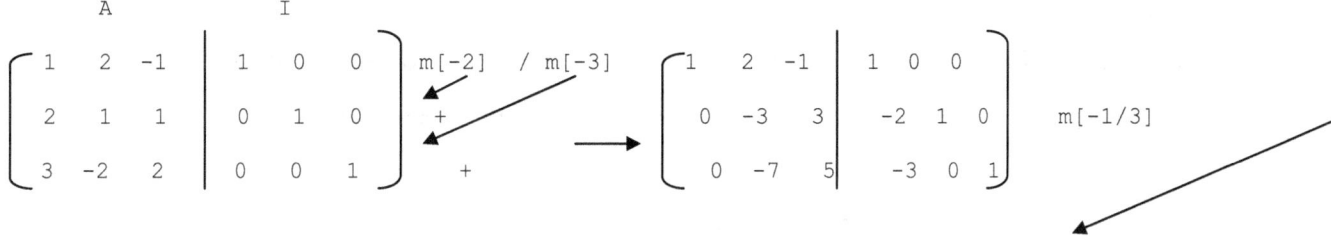

$$\begin{array}{cc} A & I \end{array}$$

$$\left[\begin{array}{ccc|ccc} 1 & 2 & -1 & 1 & 0 & 0 \\ 2 & 1 & 1 & 0 & 1 & 0 \\ 3 & -2 & 2 & 0 & 0 & 1 \end{array}\right] \begin{array}{l} m[-2] \ / \ m[-3] \\ + \\ + \end{array} \longrightarrow \left[\begin{array}{ccc|ccc} 1 & 2 & -1 & 1 & 0 & 0 \\ 0 & -3 & 3 & -2 & 1 & 0 \\ 0 & -7 & 5 & -3 & 0 & 1 \end{array}\right] \quad m[-1/3]$$

$$\begin{bmatrix} 1 & 2 & -1 & \bigm| & 1 & 0 & 0 \\ 0 & 1 & -1 & \bigm| & 2/3 & -1/3 & 0 \\ 0 & -7 & 5 & \bigm| & -3 & 0 & 1 \end{bmatrix} \begin{matrix} \\ \\ m[7] \\ + \end{matrix} \longrightarrow \begin{bmatrix} 1 & 2 & -1 & \bigm| & 1 & 0 & 0 \\ 0 & 1 & -1 & \bigm| & 2/3 & -1/3 & 0 \\ 0 & 0 & -2 & \bigm| & 5/3 & -7/3 & 1 \end{bmatrix} \begin{matrix} \\ \\ m[-1/2] \end{matrix}$$

$$\begin{bmatrix} 1 & 2 & -1 & \bigm| & 1 & 0 & 0 \\ 0 & 1 & -1 & \bigm| & 2/3 & -1/3 & 0 \\ 0 & 0 & 1 & \bigm| & -5/6 & 7/6 & -1/2 \end{bmatrix} \begin{matrix} + \\ \\ m[1] \end{matrix} / \begin{matrix} + \\ \\ m[1] \end{matrix} \longrightarrow \begin{bmatrix} 1 & 2 & 0 & \bigm| & 1/6 & 7/6 & -1/2 \\ 0 & 1 & 0 & \bigm| & -1/6 & 5/6 & -1/2 \\ 0 & 0 & 1 & \bigm| & -5/6 & 7/6 & -1/2 \end{bmatrix} +$$

$$\begin{bmatrix} 1 & 0 & 0 & \bigm| & 1/2 & -1/2 & 1/2 \\ 0 & 1 & 0 & \bigm| & -1/6 & 5/6 & -1/2 \\ 0 & 0 & 1 & \bigm| & -5/6 & 7/6 & -1/2 \end{bmatrix}$$

$$\qquad I_3 \qquad\qquad\qquad A^{-1}$$

3. check $A^{-1}$ ;

$$\begin{bmatrix} 1/2 & -1/2 & 1/2 \\ -1/6 & 5/6 & -1/2 \\ -5/6 & 7/6 & -1/2 \end{bmatrix} \cdot \begin{bmatrix} 1 & 2 & -1 \\ 2 & 1 & 1 \\ 3 & -1 & 2 \end{bmatrix} = \begin{bmatrix} 1 & 0 & 0 \\ 0 & 1 & 0 \\ 0 & 0 & 1 \end{bmatrix}$$ checks; so correct $A^{-1}$ ;

II.

1.  $A\ X\ =\ C$   gives   $X\ =\ A^{-1}\ C$

2. as in:

$$\begin{bmatrix} x_1 \\ x_2 \\ x_3 \end{bmatrix} = \begin{bmatrix} 1/2 & -1/2 & 1/2 \\ -1/6 & 5/6 & -1/2 \\ -5/6 & 7/6 & -1/2 \end{bmatrix} \cdot \begin{bmatrix} -2 \\ 1 \\ 4 \end{bmatrix} = \begin{bmatrix} 1/2 \\ -5/6 \\ 5/6 \end{bmatrix}$$

3. So, The solution set:   ( 1/2 , -5/6 , 5/6 )

4. It is of some interest to note the Geometry of this problem. Geometrically the given equations represent three straight lines in the $x_1$ $x_2$ $x_3$- space; the solution set indicates that they intersect in a single point whose coordinates are $x_1$ = 1/2 , $x_2$ = -5/6 , & $x_3$ = 5/6 in the 3-space.

III. Because of the complexity in actually finding the inverse $A^{-1}$ , we usually solve systems by the Echelon Matrix Method. We do use the concept of the Inverse in much of our theory work.

8.19 Note these Examples:

I. Solve $\begin{cases} 2x_1 - 3x_2 = 7 \\ -4x_1 + 6x_2 = 3 \end{cases}$

1. $\begin{bmatrix} 2 & -3 & 7 \\ -4 & 6 & 3 \end{bmatrix}$ m[2] $\rightarrow$ $\begin{bmatrix} 2 & -3 & 7 \\ 0 & 0 & 17 \end{bmatrix}$ $\rightarrow$ $\begin{cases} 2x_1 - 3x_2 = 7 \\ 0 = 17 \end{cases}$ not true;

so No Solution.

2. System is Inconsistent. The $2^{nd}$ equation can be multiplied be -1/2 to get the equal

form: $2x_1 - 3x_2 = -3/2$; then in one case $2x_1 - 3x_2 = 7$ and equals -3/2 in the other.

3. Note the Geometry of this problem. The given system is two lines in the $x_1\, x_2$

plane, but the "solution" shows that they are Parallel and thus do not intersect to give

a solution.

II. Solve $\begin{cases} 2x_1 - 3x_2 = 7 \\ -4x_1 + 6x_2 = -14 \end{cases}$

1. $\begin{bmatrix} 2 & -3 & 7 \\ -4 & 6 & 3 \end{bmatrix}$ m[2] $\rightarrow$ $\begin{bmatrix} 2 & -3 & 7 \\ 0 & 0 & 0 \end{bmatrix}$ $\rightarrow$ $\begin{cases} 2x_1 - 3x_2 = 7 \\ 0 = 0 \end{cases}$ true;

2. $r = 1 < 2 = n$ ; so to let $x_2$ be any number; we normally let $x_2 = t$ where t is

any number;

so $2x_1 - 3t = 7$ or $x_1 = [7 + 3t]/2$

3. Therefore the solution set [ [7 + 3t]/2 , t ] is infinite.

4 Geometrically the "two" lines in the $x_1\, x_2$ -space Coincide; so there is really

only one line. We consider that the "two" lines intersect in every point of the

line $2x_1 - 3x_2 = 7$ .

5 Note: -2 times the 1st equation gives the $2^{nd}$; so the problem as stated in the given

is a Dependent system.

III. Solve: $\begin{cases} x_1 - 4x_2 + 5x_3 = 1 \\ 2x_1 + x_2 - 3x_3 = 4 \end{cases}$

1. $\begin{bmatrix} 1 & -4 & 5 & 1 \\ 2 & 1 & -3 & 4 \end{bmatrix}$ m[-2] $\rightarrow$ $\begin{bmatrix} 1 & -4 & 5 & 1 \\ 0 & 9 & -13 & 2 \end{bmatrix}$ m[1/9] $\rightarrow$ $\begin{bmatrix} 1 & -4 & 5 & 1 \\ 0 & 1 & -13/9 & 2/13 \end{bmatrix}$

2. as a system $\left\{ \begin{array}{l} x_1 - 4x_2 + 5\ x_3 = 1 \\ x_2 - 13/9\ x_3 = 2/13 \end{array} \right\}$

3. $r = 2 < 3 = n$ : so let $x = t$ then $x_2 = [2 + 13\ t]/9$ & $x = [17 + 7\ t]/9$ .

4.  Solution set:     ( $[17 + 7t]/9$ , $[2 + 13t]/9$ , t )

5. We have two Planes in the $x_1$  $x_2$  $x_3$ -space intersecting in a Line in this 3-space. As t takes on all possible values, stept #4 gives all of the points that are on this line. We can use the methods of Anayltic Geometry to find the equation of this line if we so desire.

8.20 Homogeneous Linear Systems re-studied.

I. Consider systems of the form: A X = C where  $A = (a_{ij} : i = 1, ...,m ; j = 1, ...,n)$ ,

   $X = (x_{ij} : i = 1,... ,n ; j = 1$  , and  $C_{ij} = ( c_{ij} = 0 : i = 1, ...,m ; j = 1 )$

   1. Remember if $r = n$ then $x_1 = x_2 = x_3 = ... = x_n = 0$ is the only solution,

   2. And, if $r < n$ there are an infinite number of solutions including the above trivial
      solution; so a homogeneous system with fewer equations than the number of unknowns
      always has non-trivial solutions.

II. Theorem 8.20.1. Linear Combinations of Solutions are Solutions of Homogeneous systems.

   A. or:  If  $X' = ( x'_1 ... x'_n )$ ,  $X'' = ( x''_1 ... x''_n )$  are solutions of
the

            System  A X = 0 ,

            Then   X = $k_1$ X' + $k_2$ X''    is also a solution where the k's
                 are any constants not all zero.

   B. Proof:

      1. Given: A X' = 0    &    A X'' = 0

      2. then   A X = A [ $k_1$ X' + $k_2$ X'' ] = A $k_1$ X' + A $k_2$ X''

               = $k_1$ [A X' ] + $k_2$ [A X'' ]   = $k_1$ [0] + $k_2$ [0]  = 0 + 0

            A X = 0

      2. so this X  is a solution.

      3. this is easily expanded for the case of any number of X's not just two.

   C. Note: this is not true for the Non-Homogeneous Systems A X = C [a system where at
least one $c \ \varepsilon \ C \neq 0$.

      Proof:

1. Given:  A  X′  = C   &    A  X′′ = C

2. then   A X = A [ $k_1$ X′  + $k_2$· X′′ ] = A $k_1$ X′ + A $k_2$ X′′

   = $k_1$ [A X′ ] + $k_2$ [A X′′ ] = $k_1$ C + $k_2$ C

   = C [ $k_1$ + $k_2$ ]

which is C only for  [$k_1$ +$k_2$ ] = 1 ; so $k_1$ = 0 & $k_2$ = 1 ; or $k_1$ = 1 & $k_2$ = 0 ;

but, it is not C for any other value of the k's.

thus there is only one X as the solution.

3. this is easily expanded for the case of any number of X's not just two.

4. So, Theorem 8.20.1. is not true for non-homogeneous systems.

III.

It is clear that the only interesting Homogeneous Systems are those where

the Rank r < p when we have  ($X_1$ ,$X_2$ , ...,$X_p$ ) as the solution set as well as the

trivial solution.


IV. The Fundamental System of Solutions of Homogeneous Systems.

A.Definition: A Set of Linear Independent Solutions ( $X_1$ ,$X_2$ , ...,$X_p$ ) of the system

A X = 0 is called its Fundamental System of Solutions if every solution X can be written

as a non-trivial Linear Combination of these solutions ; or :

X = $k_1$ X + $k_2$ X + ... + $k_p$ X   where the k's are constants not all zero and

$X_i^T$ = ( $x_1$ ,$x_2$ , ...,$x_p$ ): i = 1, 2, ..., p.  [used the Transpose to save vertical

space]

B. this means:

1. The solutions form two groups. The fundamental group is all the solutions that we

can find that form an independent set. The group without a name is what is left.

2. If $k_1$ = 1 and all the other k's = 0 , then $X_1$ = X . We consider that this is

trivial and is not used in setting up the two groups.

C. See this in this example --- solve:
$$\begin{cases} x_1 + x_2 - x_3 = 0 \\ 2x_1 - x_2 - 8x_3 = 0 \\ 3x_1 + 5x_2 + x_3 = 0 \end{cases}$$

1.
$$\begin{bmatrix} 1 & 1 & -1 \\ 2 & -1 & -8 \\ 3 & 5 & 1 \end{bmatrix} \begin{matrix} m[-2]/m[-3] \\ + \\ + \end{matrix} \to \begin{bmatrix} 1 & 1 & -1 \\ 0 & -3 & -6 \\ 0 & 2 & 4 \end{bmatrix} \begin{matrix} \\ m[-1/3] \\ m[1/2] \end{matrix} \to \begin{bmatrix} 1 & 1 & -1 \\ 0 & 1 & 2 \\ 0 & 1 & 2 \end{bmatrix} m[-1] \to \begin{bmatrix} 1 & 1 & -1 \\ 0 & 1 & 2 \\ 0 & 0 & 0 \end{bmatrix}$$

2. thus r = 2 & n = 3 ; so r < n

   so there are an infinite number of solutions to

   the equivalent augmented matrix form:

$$\left\{ \begin{array}{l} x_1 + x_2 - x_3 = 0 \\ x_2 + 2x_3 = 0 \end{array} \right\}$$

3. let $x_3$ = t  then  $x_2$ = -2t  and then  $x_1$ = 3t.

4. let t = 1 then one solution is:   $X = \begin{bmatrix} 1 \\ -2 \\ 3 \end{bmatrix}$   & t = 2 gives another   $X = \begin{bmatrix} 2 \\ -4 \\ 6 \end{bmatrix}$ and   so on

5. thus, the set of solutions **X** = (t, -2t, 3t) = t(1,-2,3) = t X where X = (1,-2,3)

   do not form an independent set for they depend on each other;

   so are not a Fundamental System,

6. The Fundamental System of Solutions is one of them;

   for convenience choose X = (1,-2,3)  for the parameter.

V. Definitions:

1. Because all the solution sets ($x_1$ , $x_2$ , ..., $x_n$ ) of the homogeneous system A X = 0

   make the product of A & X zero, we call the Vector Space of these solutions

   The Null Space of A,

2. The Dimension  of this space is called The Nullity of A,

3. Note :  rank of A  + nullity of A  =  n

VI. This leads to the  Theorem 8.21.

8.21.

I.

   A.  Theorem 8.21. If the non-homogeneous system  A X = C ≠ 0  has a set of solutions

               X = ($x_1$ , $x_2$ , ..., $x_n$ ),

               then all of the solutions have the form of :  X  = $X_f$ + $X_h$   where

               $X_f$  is a fixed solution and

               $X_h$  contains all of the solutions of the homogeneous system  A X = 0.

   B. Proof:

     1. in general  A X = C  but we also have that  A $X_f$  = C

     2. subtracting:  A  X  -  A  $X_f$  = C - C

               A [ X  - $X_f$  ]  = 0

     3. thus  X  - $X_f$   is a solution of the homogeneous system  A  X  =  0

     4. so  X  -  $X_f$  = $X_h$                        or       X = $X_f$  + $X_h$   .

II. Example.   Solve:
$$\begin{cases} 3x_1 + x_2 - x_3 = 3 \\ 2x_1 + 2x_2 - 3x_3 = 1 \\ -x_1 + x_2 - 2x_3 = -2 \end{cases}$$

1. augmented matrix B:

$$\begin{bmatrix} 3 & 1 & -1 & 3 \\ 2 & 2 & -3 & 1 \\ -1 & 1 & -2 & -2 \end{bmatrix} \longrightarrow \begin{bmatrix} -1 & 1 & -2 & -2 \\ 2 & 2 & -3 & 1 \\ 3 & 1 & -1 & 3 \end{bmatrix} \begin{matrix} m[2]/m[3\backslash \\ + \\ + \end{matrix}$$

$$\begin{bmatrix} -1 & 1 & -2 & -2 \\ 0 & 4 & -7 & -3 \\ 0 & 4 & -7 & -3 \end{bmatrix} \begin{matrix} \\ m[-1] \\ + \end{matrix} \longrightarrow \begin{bmatrix} -1 & 1 & -2 & -2 \\ 0 & 4 & -7 & -3 \\ 0 & 0 & 0 & 0 \end{bmatrix} \begin{matrix} m[-1] \\ m[1/4] \\ \end{matrix} \longrightarrow \begin{bmatrix} 1 & -1 & 2 & 2 \\ 0 & 1 & -7/4 & -3/4 \\ 0 & 0 & 0 & 0 \end{bmatrix}$$

2.  $r = 2 < 3 = n$  ; so there is  an infinite set of solutions.

3. the  A X = C equations:
$$\begin{cases} -x_1 + x_2 - 2x_3 = -2 \\ 4x_2 - 7x_3 = -3 \end{cases}$$

4.  let  $x_3 = t$   then  $x_2 = [7t - 3]/4$   and   $x_1 = [5 - t]/4$

5.  let  t = 1  to get:
$$x_f = \begin{bmatrix} 1 \\ 1 \\ 1 \end{bmatrix}$$

6. the homogeneous equations A X = 0:
$$\begin{cases} -x_1 + x_2 - 2x_3 = 0 \\ 4x_2 - 7x_3 = 0 \end{cases}$$

7.  let  $x_3 = s$   then  $x_2 = 7s/4$   and  $x_1 = -s/4$

8.  thus
$$x_h = \begin{bmatrix} -s/4 \\ 7s/4 \\ s \end{bmatrix}$$

9.

All solutions  X =
$$\begin{bmatrix} 1 - s/4 \\ 1 + 7s/4 \\ 1 + s \end{bmatrix}$$
s  any real number.

VII. Theorem.  Given:  the mxn System of Linear Equations   $A X = b$ where

$$A = (a_{ij})_{mxn} \quad , \quad X = (x_{ij})_{mxn} \quad , \quad \& \quad b = (b_{ij})_{mx1}$$

OR $\qquad A = \begin{bmatrix} a_{11} & \cdots & a_{1n} \\ \cdots & \cdots & \cdots \\ \cdots & \cdots & \cdots \\ a_{m1} & \cdots & a_{mn} \end{bmatrix} \qquad \& \qquad b = \begin{bmatrix} b_1 \\ \cdot \\ \cdot \\ b_m \end{bmatrix}$

let  B  be the Augement Matrix of the System; as:

$$B = \begin{bmatrix} a_{11} & \cdots & a_{1n} & b_1 \\ \cdots & \cdots & \cdots & \cdot \\ \cdots & \cdots & \cdots & \cdot \\ a_{m1} & \cdots & a_{mn} & b_m \end{bmatrix}$$

Then A  &  B  have the Same Rank.

Proof.

   I.  Let the  Rank of  B  be  r :

      1. change B to its echelon form; this form of B has the same rank, r ,

      2. there are  r  non-zero rows in  B  where at least one of the non-zero

         rows has at least one non-zero $a_{ij}$,

      2. A  is the same as  B  without the  b  column ,

      3. dropping the b's can not effect the number of non-zero rows ,

      4. the Rank of A is  r ,

      5. so:  if the Rank of B is  r , then the Rank of A is  r.

  II. Let the Rank of A  be  r  :

      1. from a previous theorem the Column Rank of A is  r ,

      2. so A has  r  Indepentent Columns ,

      3. given  A X = b  then  b is determined by a Linear Combination

         of the column vectors of A ; as : $C_1 \ x_1 \ + C_2 \ x_2 \ + \ldots + C_n \ x_n \ = \ b$

         where $C_k = (a_{ij})_{mxk}$ ,

      4. change A  into  B by adding the b  column at the end of the columns,

      5. because this adds no new independent columns , the Rank is still  r ,

      6. so:  if the Rank of A is  r  ,  then the Rank of B is  r.

III.   Therefore the Ranks of A  & B are the same.  We see that in all of our work with Rank in Systems of Linear Equations, we can use either the A Matrix  or  the B Matrix.

8.22 Exercises - find All Solutions of:

1.
$$\begin{cases} 2x_1 & + & 2x_2 & - & x_3 & + & x_4 & = & -2 \\ 3x_1 & - & x_2 & + & 3x_3 & & & = & 0 \\ & & 3x_2 & - & 2x_3 & - & x_4 & = & 9 \\ x_1 & & & + & x_3 & - & 2x_4 & = & 9 \end{cases}$$
$[-1,3,2,-4]$

2.
$$\begin{cases} 2x_1 & + & x_2 & + & 3x_3 & + & 4x_4 & = & 0 \\ 3x_1 & + & 4x_2 & + & 7x_3 & + & 6x_4 & = & 0 \\ x_1 & & & + & x_3 & + & 2x_4 & = & 0 \end{cases}$$
$[-t-2s,-t,t,s]$

## 9.1 Introduction.

I. We wish to look at the System of n Linear Equations in n Unknowns in a different light than in the previous chapter. We will develop a method that pre-dates the matrix method, but we will update this method by referring to matrices when convenient.  It is not a very good method for calculations when the order is large, but it introduces us to a concept that has real uses in extended theoretical work in Linear Algebra and even in the Calculus. The concept is loosely covered in most elementary courses so is not new, but we do need to use the concept in later Linear Algebra studies. It seems to fit here.

II. The development of this new concept from a simple thing  to a general form is complicated. There are two approaches that can be used; we will use the one that we  feel is the simplest of the two. Even here we will not try to give a complete coverage of the theory; the algebra can get very involved. We will proceed through the concept in a manner that is easy to understand; a manner that gives us all the information that we need for this text. We will give meaning and simple proofs to most of  what we do.

III. Consider a 2x2 system of equations:

1. Given $\begin{cases} a_{11} \ x_1 \ + \ a_{12} \ x_2 \ = \ c_1 \quad [\alpha] \\ a_{21} \ x_1 \ + \ a_{22} \ x_2 \ = \ c_2 \quad [\beta] \end{cases}$ note the matrix formats $\begin{bmatrix} a_{11} & a_{12} \\ a_{21} & a_{22} \end{bmatrix}$ & $\begin{bmatrix} c_1 \\ c_2 \end{bmatrix}$

2. solve  by:  $m[\alpha]$ by $a_{22}$ :      $a_{11} \ a_{22} \ x_1 \ + \ a_{12} \ a_{22} \ x_2 \ = \ a_{22} \ c_1$

   & $m[\beta]$ by $-a_{12}$ :      $-a_{21} \ a_{12} \ x_1 \ - \ a_{22} \ a_{12} \ x_2 \ = \ -a_{12} \ c_2$

3. add:                      $[a_{11} \ a_{22} \ - \ a_{21} \ a_{12} \ ]x_1 \ = \ a_{22} \ c_1 \ - \ a_{12} \ c_2$

4. so:
$$x_1 \ = \ \frac{a_{22} \ c_1 \ - \ a_{12} \ c_2}{a_{11} \ a_{22} \ - \ a_{21} \ a_{12}}$$

5. with this method we find that
$$x_2 \ = \ \frac{a_{21} \ c_1 \ - \ a_{11} \ c_2}{a_{11} \ a_{22} \ - \ a_{21} \ a_{12}}$$

Note: we could find $x_2$  by substitution instead but this gets quite messy.

6. one could note that diagonal products on the array $\begin{matrix} a_{11} & a_{12} \\ a_{21} & a_{22} \end{matrix}$ would

$\qquad\qquad - \qquad\qquad +$

give us the denominators. We could get the numerators with such products

if the ith column is replaced with the constant column when solving for $x_i$.

IV.

1. A matrix is not a number; so the array used in this manner has to be given a new

name, a new definition, a new symbol, and different theorems, properties, etc. We

will develop this new concept using the system of equations and its solution as a

guide.

2. While a matrix does not represent a number , many times it is  useful to assign

a number to some special computation involving the elements of a matrix

--- computation as in III. 6. This very important number is called the Determinant

of a Matrix. Only Square Matrices have Determinants.

3. This is the same concept that students have studied in previous elementary courses in

a more informal manner, but we will start over from the beginning.

9.2.   I.

1. Given a Square Matrix A where A is written as

$$A = \begin{bmatrix} a_{11} & a_{12} \\ \\ a_{21} & a_{22} \end{bmatrix}$$

then the Determinant of A is written as

$$|A| = \begin{vmatrix} a_{11} & a_{12} \\ \\ a_{21} & a_{22} \end{vmatrix}$$

and symbolized by  det$|A|$   or   $|A|$

2. We can increase the size [ called Order ] of the Determinant by increasing the

number of the rows/columns equally, but it must always be square.

II.   Definitions --- based on the ideas of 9.1.

A. It seems reasonable to include: a 1st Order Determinant is defined by $|A| = |a_{11}| = a_{11}$

B. A 2nd Order Determinant.

1. Is defined by:

$$|A| = \begin{vmatrix} a_{11} & a_{12} \\ \\ a_{21} & a_{22} \end{vmatrix} = a_{11} \ a_{22} \ - \ a_{12} \ a_{21}$$

2. We will use the set of two diagonal products to get the number that is

the value of the determinant:

from        $a_{11}$        $a_{12}$              we get the      $a_{11}$    $a_{22}$    $-$    $a_{12}$    $a_{21}$

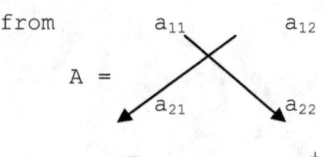

A =

        $a_{21}$        $a_{22}$

        $-$            $+$

  C. we can indicate the solution to our system in a simple form involving these $2^{nd}$

order determinants; [note that the expansions agree with 9.1.III.4 & 5.]:

$$x_1 = \frac{\begin{vmatrix} c_1 & a_{12} \\ c_2 & a_{22} \end{vmatrix}}{\begin{vmatrix} a_{11} & a_{12} \\ a_{21} & a_{22} \end{vmatrix}} \qquad x_2 = \frac{\begin{vmatrix} a_{11} & c_1 \\ a_{21} & c_2 \end{vmatrix}}{\begin{vmatrix} a_{11} & a_{12} \\ a_{21} & a_{22} \end{vmatrix}}$$

III. Now consider a 3x3 system of equations.

A. Given $\begin{cases} a_{11} \; x_1 + a_{12} \; x_2 + a_{13} \; x_3 = c_1 \\ a_{21} \; x_1 + a_{22} \; x_2 + a_{23} \; x_3 = c_2 \\ a_{31} \; x_1 + a_{32} \; x_2 + a_{33} \; x_3 = c_3 \end{cases}$  $\begin{matrix}[\alpha] \\ [\beta] \\ [\delta]\end{matrix}$

B.

  1. Use the method of 9.1. III. to eliminate $x_3$ from $[\alpha]\&[\beta]$ and from $[\alpha]\&[\delta]$ ;

     this gives an equivalent $2^{nd}$ order system in just $x_1$ & $x_2$ ;

  2. eliminate $x_2$ in this system to obtain an equation in just $x_1$ ;

  3. this $x_1$ will look like:

$$x_1 = \frac{a_{12}\,a_{23}\,c_3 + a_{13}\,c_2\,a_{32} + c_1\,a_{33}\,a_{22} - c_1\,a_{23}\,a_{32} - a_{12}\,c_2\,a_{33} - a_{22}\,a_{13}\,c_3}{a_{11}\,a_{22}\,a_{33} + a_{12}\,a_{23}\,a_{31} + a_{13}\,a_{21}\,a_{32} - a_{13}\,a_{22}\,a_{31} - a_{11}\,a_{23}\,a_{32} - a_{12}\,a_{21}\,a_{33}}$$

     which is a solution,

  4. we would find that $x_2$ & $x_3$ have similar forms;

C. Because of the very complicated form for the evaluation of each x work has been done

to discover an elegant device to illustrate each solution and a method to convert each

such form into a specific solution for each x:

1.

$$x_1 = \frac{\begin{vmatrix} c_1 & a_{12} & a_{13} \\ c_2 & a_{22} & a_{23} \\ c_3 & a_{32} & a_{33} \end{vmatrix}}{\begin{vmatrix} a_{11} & a_{12} & a_{13} \\ a_{21} & a_{22} & a_{23} \\ a_{31} & a_{32} & a_{33} \end{vmatrix}} \qquad x_2 = \frac{\begin{vmatrix} a_{11} & c_1 & a_{13} \\ a_{21} & c_2 & a_{23} \\ a_{31} & c_3 & a_{33} \end{vmatrix}}{\begin{vmatrix} a_{11} & a_{12} & a_{13} \\ a_{21} & a_{22} & a_{23} \\ a_{31} & a_{32} & a_{33} \end{vmatrix}}$$

$$x_3 = \frac{\begin{vmatrix} a_{11} & a_{12} & c_1 \\ a_{21} & a_{22} & c_2 \\ a_{31} & a_{32} & c_3 \end{vmatrix}}{\begin{vmatrix} a_{11} & a_{12} & a_{13} \\ a_{21} & a_{22} & a_{23} \\ a_{31} & a_{32} & a_{33} \end{vmatrix}}$$

2. we will define the $3^{rd}$ order Determinants to fit the denominators of B.3. :

$$|A| = \begin{vmatrix} a_{11} & a_{12} & a_{13} \\ a_{21} & a_{22} & a_{23} \\ a_{31} & a_{32} & a_{33} \end{vmatrix}$$

$$= \underset{1}{a_{11}\,a_{22}\,a_{33}} + \underset{2}{a_{12}\,a_{23}\,a_{31}} + \underset{3}{a_{13}\,a_{21}\,a_{32}} - \underset{4}{a_{13}\,a_{22}\,a_{31}} - \underset{5}{a_{11}\,a_{23}\,a_{32}} - \underset{6}{a_{12}\,a_{21}\,a_{33}} \;;$$

these represent the number of each product indicated in the mnemonic device below:

3. one graphical mnemonic that does give the correct number value of this $|A|$: involves

   duplicating the $1^{st}$ & $2^{nd}$ columns as extra $4^{th}$ & $5^{th}$ columns:

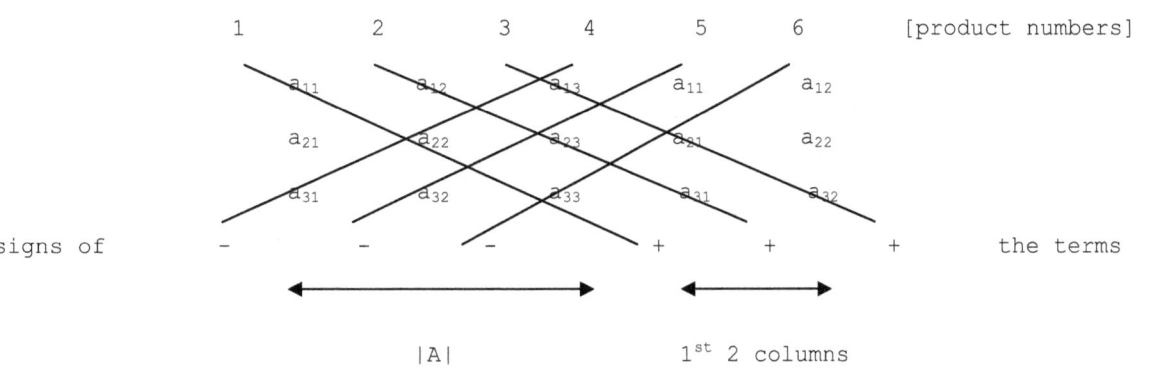

IV.

1. Using the method of III. B. and an enormous amount of work it is possible to arrive at a definition of a $4^{th}$ order, but it is almost an impossible job to do this for a $5^{th}$ order. In addition it is not possible to find a diagonal method for evaluating $4^{th}$ and higher order determinants; so the above method fails for $n > 4$.

2. We will use a different approach for a general definition by starting with two new definitions.

9.3  Definitions.

A.

1. The Minor M    of the $a_{ik}$ th  element of a determinant is the determinant left after deleting the ith row and the kth column of the original determinant.

2. as—given:
$$|A| = \begin{vmatrix} a_{11} & a_{12} & a_{13} \\ a_{21} & a_{22} & a_{23} \\ a_{31} & a_{32} & a_{33} \end{vmatrix} \qquad \text{then} \qquad M_{23} = \begin{vmatrix} a_{11} & a_{12} \\ a_{31} & a_{32} \end{vmatrix}$$

B.

1. The Cofactor $C_{ik}$    of the $a_{ik}$ th   element of  a determinant is  defined as $[-1]^{i+k}$ times the Minor of that element ; as:   $C_{ik}$    $= [-1]^{i+k} M_{ik}$      ;

2. using the example:   $C_{23}$   $= [-1]^{2+3} \begin{vmatrix} a_{11} & a_{12} \\ a_{31} & a_{32} \end{vmatrix} = [-1]^5 [a_{11} a_{32} - a_{12} a_{31}]$

or             $C_{23}$   $= -a_{11} a_{32} + a_{12} a_{31}$ .

A. We start by looking at evaluating order 3 in another manner:   now

$$|A| \quad = \quad \begin{vmatrix} a_{11} & a_{12} & a_{13} \\ a_{21} & a_{22} & a_{23} \\ a_{31} & a_{32} & a_{33} \end{vmatrix}$$

$$= \quad a_{11}\,a_{22}\,a_{33} + a_{12}\,a_{23}\,a_{31} + a_{13}\,a_{21}\,a_{32} - a_{13}\,a_{22}\,a_{31} - a_{11}\,a_{23}\,a_{32} - a_{12}\,a_{21}\,a_{33}$$

B.

1. Because of the Commuative Law we could write the terms in a different order: as

$$|A| = - a_{12}\,a_{21}\,a_{33} + a_{13}\,a_{21}\,a_{32} + a_{11}\,a_{22}\,a_{33} - a_{13}\,a_{22}\,a_{31} - a_{11}\,a_{23}\,a_{32} + a_{12}\,a_{23}\,a_{31}$$

the      6$^{th}$              3$^{rd}$           1$^{st}$              4$^{th}$              5$^{th}$              2$^{nd}$    terms of A.

2. grouping in pairs and factoring out the common element we get:

$$|A| = - a_{21}\,[a_{12}\,a_{33} - a_{13}\,a_{32}] + a_{22}\,[a_{11}\,a_{33} - a_{13}\,a_{31}] - a_{23}\,[a_{11}\,a_{32} + a_{12}\,a_{31}]$$

3. note:

$$|A| = [-1]^{2+1}\,a_{21}\begin{vmatrix} a_{12} & a_{13} \\ a_{32} & a_{33} \end{vmatrix} + [-1]^{2+2}\,a_{22}\begin{vmatrix} a_{11} & a_{13} \\ a_{31} & a_{33} \end{vmatrix} + [-1]^{2+3}\,a_{23}\begin{vmatrix} a_{11} & a_{12} \\ a_{31} & a_{32} \end{vmatrix}$$

$$= \quad [-1]^{2+1}\,a_{21}\,M_{21} + [-1]^{2+2}\,a_{22}\,M_{22} + [-1]^{2+3}\,a_{23}\,M_{23}$$

$$|A| = \quad a_{21}\,C_{21} + a_{22}\,C_{22} + a_{23}\,C_{23} \qquad ;$$

4. Or, the determinant |A| can be evaluated by: in the 2$^{nd}$ row of A find the Sum of the Products of the elements and their Cofactors.

5. By changing the order and the grouping of Steps 1 - 2 we can get:

$$|A| = \quad a_{11}\,C_{11} + a_{12}\,C_{12} + a_{13}\,C_{13} \qquad ;$$

C. It is clear that we can show that this can be done using any row [ or column ] and with any order where n ≥ 3. The fact that this can be done for all determinants [ n > 2 ] is proof enough by induction that this is the definition/value that we should use for all such determinants. This is called Evaluating a Determinant by Cofactors. The method cuts an nth order down to a series of [n −1] orders; we continue until each determinant left is of the 3$^{rd}$ or 2$^{nd}$ order and finish the evaluation by using the diagonal products. In 9.6 we will improve on this method.

9.5. Here is a list of Properties of Determinants actually used in computations.

These are like little theorems or things to do. Because of the complexity of n>3 cases,

we only will look at specific or general $2^{nd}$ or $3^{rd}$ order determinants to check or

prove their truths. For this text these restrictions of n = 2 or n = 3 will serve

most of our needs. We will refer to the Elementary Row Operations on Matrices as

Types I. , II. , & III. and handle our theorems in this manner: given a matrix A perform

Type I. [or II. / III.] to get matrix B; then find the determinants |A| & |B|; then

compare |A| & |B|.

A. TypeI: get B by interchanging two rows in A; then    |B| = - |A|.

   1. let   $|A| = \begin{vmatrix} a & b \\ c & d \end{vmatrix} = ad - bc$   and   $|B| = \begin{vmatrix} c & d \\ a & b \end{vmatrix} = bc - ad = - [ ad - bc ]$

   2. so  |B| = - |A|      note: the difference compared to matrices.

   3. Exercise: student verify this for a general $3^{rd}$ order determinant.


B. Any constant factor of a row  may [ and should ] be factored out to the outside of the

determinant to serve as its first factor  -- in matrices these factors

 were thrown away but not here. This procedure uses Type II: get B by multiplying a

row by a constant k ; then  |B| = k |A|.

   1. let   $|A| = \begin{vmatrix} a & b \\ c & d \end{vmatrix} = ad - bc$   and  $|B| = \begin{vmatrix} a & b \\ kc & kd \end{vmatrix} = kad - kbc = k [ad - bc]$

   2. so  |B| = k |A|

   3. later we will go into more detail on this concept.


C. A row or column of a determinant |A|  [called a Secondary] can be changed or augmented

by multiplying each element of some other row or column of |A|  [called the Pivotal or

Primary row or column ] by a constant k and adding it to the corresponding elements

of the secondary [the pivotal row or column does not change] to get a determinant |B| ;

then |B| = |A| . This is the type III.

1. let $|A| = \begin{vmatrix} a & b \\ c & d \end{vmatrix} = ad - bc$

2. and $|B| = \begin{vmatrix} a & b \\ ka+c & kb+d \end{vmatrix} = kab + ad - kab - bc = ad + bc = |A|$

3. Exercise: student verify this for a general $3^{rd}$ order determinant.


## 9.6. Evaluating Determinants.

1. Apply these properties over and over again so that all the secondaries have 0's in the same row [or column] - except for the element in the row [or column] of the pivotal column [or row]. This non-zero element is the Pivotal Element and is what one uses to get the 0's.

2. Expand by the cofactors of that row [ or column ] to evaluate the determinant. Except for the pivotal element all the elements in that row [ or column ] will be 0's. There will be left a single, non-zero determinant of one less order. This determinant is equal in value to the original. Using fraction and negatives if necessary it is always possible to do this.

3. Continue this process until the last determinant is of the $2^{nd}$ order which is then easily evaluated.

4. Best choice for the pivotal element is a 1, and one may want to use the Type II. operation to get this 1. For what happens later it is probably best to use the last column to get the 0's in the row of the pivotal element. Note: this skill has already been developed when we got echelon forms with matrices.

9.7 Example. Evaluate:

$$\begin{array}{cccc} + & - & + & - \end{array} \quad \text{[the sign positions in the } 1^{st} \text{ row]}$$

$$\begin{vmatrix} 2 & -1 & 2 & -1 \\ 1 & 1 & -1 & 2 \\ 1 & 2 & 1 & -3 \\ 1 & 3 & 4 & 7 \end{vmatrix} = \begin{vmatrix} 0 & 0 & 0 & -1 \\ 5 & -1 & 3 & 2 \\ -5 & 5 & -5 & -3 \\ 15 & -4 & 18 & 7 \end{vmatrix} = -[-1]\begin{vmatrix} 5 & -1 & 3 \\ -5 & 5 & -5 \\ 15 & -4 & 18 \end{vmatrix} = +[-5]\begin{vmatrix} 5 & -1 & 3 \\ 1 & -1 & 1 \\ 15 & -4 & 18 \end{vmatrix}$$

m[2]

m[-1]

m[2]

m[1]

m[-1]

$$= -5 \begin{vmatrix} 2 & 2 & 3 \\ 0 & 0 & 1 \\ -3 & 14 & 18 \end{vmatrix} = -[-5][1] \begin{vmatrix} 2 & 2 \\ -3 & 14 \end{vmatrix} = 5 [ 28 + 6 ] = 170$$

9.8 Cramer's Rule for solving a System of n Simultaneous Linear Equations in n Unknowns.

1. This rule is one application of the use of determinants, and it can be used with only the theory of determinants that we have studied so far. It has its main use in theory , because the arithmetic gets too complicated for n > 3. We still use the rule because most of our work will be with n < 4.

2. The Rule. Solve the nxn system of $1^{st}$ degree equations where n > 2 ; so

given:
$$\begin{cases} a_{11} \ x_1 + \ldots + a_{1n} \ x_n = c_1 \\ \ldots \ldots \ldots \ldots \ldots \ldots \ldots \ldots \\ \ldots \ldots \ldots \ldots \ldots \ldots \ldots \ldots \\ a_{n1} \ x_1 + \ldots + a_{nn} \ x_n = c_n \end{cases}$$

assume rank r of A [ $a_{ij}$ ]$_{nxn}$    is  n ,

then  X   =   |$A_i$ | / |A|     for  i = 1, 2, ...,n

where

$$|A| = \begin{vmatrix} a_{11} \ldots \ldots \ldots a_{1n} \\ \ldots \ldots \ldots \ldots \\ \ldots \ldots \ldots \ldots \\ a_{n1} \ldots \ldots \ldots a_{nn} \end{vmatrix}$$  the determinant of the coefficients

and    |A | is  |A| with the ith column replaced with  $c_1$   or constant column.

3. The proof is just a manner of solving for the x's by algebra and showing that the solutions fit the above forms of  |A| & |$A_i$ |. We have no interest in the proof ---just the example in 9.9.

9.9 Solve by Cramer's Rule :
$$\begin{cases} 2x_1 + 2x_2 - x_3 + x_4 = -2 \\ 3x_1 - x_2 + 3x_3 \qquad = 0 \\ \qquad 3x_2 - 2x_3 - x_4 = 9 \\ x_1 \qquad + x_3 - 2x_4 = 9 \end{cases}$$

1.

$$|A| = \begin{vmatrix} 2 & 2 & -1 & 1 \\ 3 & -1 & 3 & 0 \\ 0 & 3 & -2 & -1 \\ 1 & 0 & 1 & -2 \end{vmatrix} = \begin{vmatrix} 0 & 0 & 0 & 1 \\ 3 & -1 & 3 & 0 \\ 2 & 5 & -3 & -1 \\ 5 & 4 & -1 & -2 \end{vmatrix} = -[1] \begin{vmatrix} 3 & -1 & 3 \\ 2 & 5 & -3 \\ 5 & 4 & -1 \end{vmatrix} = - \begin{vmatrix} 18 & 11 & 3 \\ -13 & -7 & -3 \\ 0 & 0 & -1 \end{vmatrix}$$

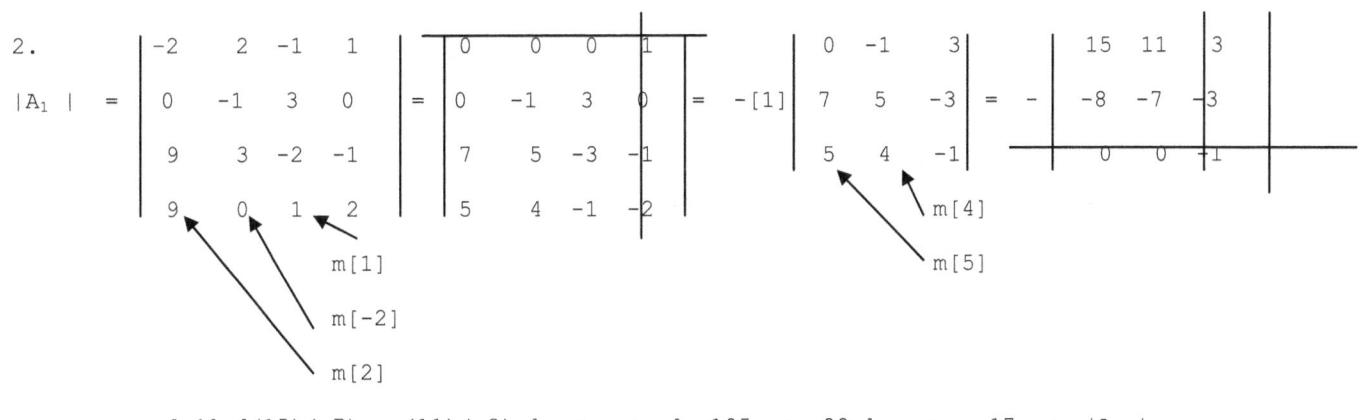

m[1]

m[-2]

m[-2]

m[4]

m[5]

$$= -[-1] \quad \begin{array}{cc} 18 & 11 \\ -13 & -7 \end{array} \quad = \quad + [ -126 + 143 ] = 17 = |A|$$

2.

$$|A_1| = \begin{vmatrix} -2 & 2 & -1 & 1 \\ 0 & -1 & 3 & 0 \\ 9 & 3 & -2 & -1 \\ 9 & 0 & 1 & 2 \end{vmatrix} = \begin{vmatrix} 0 & 0 & 0 & 1 \\ 0 & -1 & 3 & 0 \\ 7 & 5 & -3 & -1 \\ 5 & 4 & -1 & -2 \end{vmatrix} = -[1] \begin{vmatrix} 0 & -1 & 3 \\ 7 & 5 & -3 \\ 5 & 4 & -1 \end{vmatrix} = - \begin{vmatrix} 15 & 11 & 3 \\ -8 & -7 & -3 \\ 0 & 0 & -1 \end{vmatrix}$$

m[1]

m[-2]

m[2]

m[4]

m[5]

$$= -[-1] [ (15)(-7) - (11)(-8) ] = + [ -105 + 88 ] = - 17 = |A_1|$$

3. In the same fashion $|A_2| = 51$ & $|A_3| = 34$

4. now we could find $|A|$ in the same manner and then find the x 's by $|A_1| / |A|$,

but it is simplier to find $x_1 = |A_1| / |A|$ , $x_2 = |A_2| / |A|$ , $x_3 = |A_3| / |A|$ ,

and then find $x_4$ by substituting the other x's into the most convenient equation

for this purpose.

5. as : $x_1 = -17/17 = -1$ & $x_2 = 51/17 = 3$ & $x_3 = 34/17 = 2$ , and then into

the 4th equation we get: $-1 + 2 - 2x = 9$ => $x_4 = -4$

6. we check these results by substituting into all four equations to get

$-2 = -2$ & $0 = 0$ & $9 = 9$ & $9 = 9$.

I. Here we wish to add to the material of 9.1 a little without expanding the concept a

significant amount. We will not prove most of the cases although we may indicate the

general idea of a proof. . Some of the statements involve using the elementary

operations; so they are obviously true just by inspection. With  others we can test the

statement in a specific $3^{rd}$ order and then see that it would be true for all $3^{rd}$ orders.

The  higher orders would then seem to be true by induction. However, some others require

concepts that would take us further than we wish to go in this text; so their proofs

will be ignored. We leave it to the student to look at examples of their choice and be

convinced that the  statements are true.

II. The Homogeneous system has more than the trivial solution iff $|A| = 0$.

  A. The proof for the case of a 2x2 system is interesting ; so we will look at that.

  B. take at the system:
$$\begin{cases} ax + bx = 0 \\ cx + dx = 0 \end{cases}$$

    1. $|A| \equiv \begin{vmatrix} a & b \\ c & d \end{vmatrix} \equiv ad - bc$ ; now assume this $ad - bc = 0$

    2. $A = \begin{bmatrix} a & b \\ c & d \end{bmatrix} \xrightarrow{m[-c/a]} \begin{bmatrix} a & b \\ 0 & -bc/a + d \end{bmatrix} \rightarrow \begin{bmatrix} a & b \\ 0 & [ad - bc]/a \end{bmatrix} \rightarrow \begin{bmatrix} a & b \\ 0 & 0 \end{bmatrix}$

    3. $r = 1 < n$ ; so there are solutions other than the trivial one.

  C. We can also say that a Non-Homogeneous system has one and only one solution set

iff $|A| = 0$. Here $r = n$ ; so the last step of the echelon form ensures a single solution.

III. Others.

  1. The value of a determinant is not changed if its rows are written as columns in the

same order; as: $1^{st}$ row to become the $1^{st}$ column, etc. Or,  $|A| = |A^T|$. All the methods

of evaluation would involve the same arithmetic whether they came from $|A|$ or $|A^T|$.

  2.If all the elements of a row [ or column ] are zero, then the determinant is zero.

Every product or term would have a zero factor.

  3. If two rows [ or columns ] are the same, element by element, then the determinant is

zero --- we see this sometimes after a common factor is factored out of a row [ or

column ]. If we interchange the two identical rows [or columns], we get the same

determinant with a minus sign; or $|A| = -|A|$  =>  $|A| + |A| = 0$  =>  $|A| = 0$.

4. |A B| = |A| |B|. The proof for the general case of  nxn A & B is more involved than we wish to go into , but we will look at a proof for 2x2 matrices:

a. let A = $\begin{bmatrix} a & b \\ c & d \end{bmatrix}$ ;so |A| = ad – bc  and let B = $\begin{bmatrix} u & w \\ x & y \end{bmatrix}$ ; so |B| = uy – wx

b.

| | | u | w |
|---|---|---|---|
| | | x | y |
| A  B  = | a  b | [au + bx] | [aw + by] |
| | C  d | [cu + dx] | [cw + dy] |

c. |A B| = $\begin{vmatrix} [au + bx] & [aw + by] \\ [cu + dx] & [cw + dy] \end{vmatrix}$

d. |A B| = [au + bx][cw + dy] – [aw + by][cu + dx]

= aduy – adxw – bcyu + bcxw

d. while |A| |B| = [ad – bc] [uy – wx] = aduy – adxw –bcuy + bcxw

e. thus  |A B| = |A| |B|.

6. The I matrix has 1's along the principal diagonal and 0's every where else, and the evaluation of |I| would involve the product of those 1's for one term and have one of those 0's as a factor in every other term. Thus |I| = 1

9.11 Finding Rank by Determinants of the matrix A = $[a_{ij}]_{m \times n}$ .

1. mentally list all  of the square submatrices in a largest to smaller order;

2. find their determinants starting with the largest;

3. stop this procedure on the first non-zero value obtained;

4. the Order of this first non-zero determinant is the Rank of the A by a theorem that we will not prove in this text. Some writers use this as the basic Definition of Rank of a Matrix A to avoid the proof.

5. when finding rank by this method in some problems it is well to block off the submatrix used to find this rank. It is clear that there is much work for n > 3.

9.12. Solving systems of Linear Equations.

I. When solving a mxn system , we can find rank by the Echelon form, by Determinants, or a mixture of the two. Then, we can solve the system by the echelon form or by Cramer's Rule.

III. Solve the system:

$$\begin{cases} x_1 + 2x_2 - x_3 - x_4 - 2x_5 = 3 \\ 2x_1 + x_2 - 3x_3 - 3x_4 + 2x_5 = 4 \\ -2x_1 + 2x_2 + x_3 + x_4 + x_5 = 1 \\ x_1 + 5x_2 - 3x_3 + 3x_4 + x_5 = 8 \end{cases}$$

1.
$$\begin{bmatrix} 1 & 2 & -1 & -1 & -2 & 3 \\ 2 & 1 & -3 & -3 & 2 & 4 \\ -2 & 2 & 1 & 1 & 1 & 1 \\ 1 & 5 & -3 & 3 & 1 & 8 \end{bmatrix} \text{m[-2]/m[2]/m[-1]} \begin{bmatrix} 1 & 2 & -1 & -1 & -2 & 3 \\ 0 & -3 & -1 & -1 & 6 & -2 \\ 0 & 6 & -1 & -1 & -3 & 7 \\ 0 & 3 & -2 & 4 & 3 & 5 \end{bmatrix} \text{m[2]/m[1]}$$

$$\begin{bmatrix} 1 & 2 & -1 & -1 & -2 & 3 \\ 0 & -3 & -1 & -1 & 6 & -2 \\ 0 & 0 & -3 & -3 & 9 & 3 \\ 0 & 0 & -3 & 3 & 9 & 3 \end{bmatrix} \begin{matrix} \\ \\ \text{m[-1/3]} \\ \text{m[-1/3]} \end{matrix} \longrightarrow \begin{bmatrix} 1 & 2 & -1 & -1 & -2 & 3 \\ 0 & -3 & -1 & -1 & 6 & -2 \\ 0 & 0 & 1 & 1 & -3 & -1 \\ 0 & 0 & 1 & -1 & -3 & -1 \end{bmatrix} \text{m[-1]}$$

$$\begin{bmatrix} 1 & 2 & -1 & -1 & -2 & 3 \\ 0 & -3 & -1 & -1 & 6 & -2 \\ 0 & 0 & 1 & 1 & -3 & -1 \\ 0 & 0 & 0 & -2 & 0 & 0 \end{bmatrix}$$

change to a system of equations form

2.
$$\begin{cases} x_1 + 2x_2 - x_3 - x_4 - 2x_5 = 3 \\ -3x_2 - x_3 - x_4 + 6x_5 = -2 \\ x_3 + x_4 - 3x_5 = -1 \\ -2x_4 = 0 \end{cases}$$

it is clear that $x_4 = 0$ ; so
the system is now:

3.
$$\begin{cases} x_1 + 2x_2 - x_3 - 2x_5 = 3 \\ -3x_2 - x_3 + 6x_5 = -2 \\ x_3 - 3x_5 = -1 \end{cases}$$

4.  Augemented matrix :
$$\begin{bmatrix} 1 & 2 & -1 & -2 & 3 \\ 0 & -3 & -1 & 6 & -2 \\ 0 & 0 & 1 & -3 & -1 \end{bmatrix}$$

so Rank r = 3  where n = 4 ; thus one x is arbitrary. Let $x_5 = \alpha$

5. the new system will be:   .
$$\left\{ \begin{array}{l} x_1 + 2x_2 - x_3 = 3 + 2\alpha \\ \quad\quad -3x_2 - x_3 = -2 - 6\alpha \\ \quad\quad\quad\quad x_3 = -1 + 3\alpha \end{array} \right\}$$

so:                                          $x_3 = -1 + 3\alpha$

this $x_3$ into the 2$^{nd}$ equation:              $-3x_2 = -3 - 3\alpha$

  or                                        $x_2 = 1 + \alpha$

and  $x_2$ & $x_3$ into the first equation:     $x_1 = 3 + 2\alpha - 2 - 2\alpha - 1 + 3\alpha$

  or                                        $x_1 = 3\alpha$

6. Solution Set  [ $x_1 = 3\alpha$, $x_2 = 1 + \alpha$, $x_3 = -1 + 3\alpha$, $x_4 = 0$, $x_5 = \alpha$ ]

9.13. Determinants and  Inverse Matrices --- Given  $A = [a_{ij}]_{n \times n}$     :

I. A is  Non-Singular [ $A^{-1}$   exists ] iff $|A| \neq 0$.

   1. $A A^{-1} = I$ ; so $|A A^{-1}| = |I|$ ; thus $|A| |A^{-1}| = 1$.

   2. Now determinants are numbers ; so by algebra $|A^{-1}| = 1/ |A|$ iff $|A| \neq 0$;

   3. thus A is Non-singular for it exists.

II.  A is Singular if $|A| = 0$.

   1. given  $|A| = 0$

   2. assume A is non-singular; then $|A| \neq 0$

   3. contradiction ; therefore A is Singular.

III. det $A \cdot$ det $A^{-1} = 1$ iff  det $A \neq 0$ &  det $A^{-1} \neq 0$

   1. now from before  det $[A B]$ = det $A$  $\cdot$ det $B$

   2. so    det $[A A^{-1}]$ = det $A \cdot$  det $A^{-1}$

   3. but  $A A^{-1} = I$ then  det $I$ = det $A$  $\cdot$ det $A^{-1}$ iff  det $A \neq 0$  &  det $A^{-1} \neq 0$

   4. det $I = 1$ thus det $A$  $\cdot$  det $A^{-1} = 1$  iff  det $A \neq 0$  &  det $A^{-1} \neq 0$

IV. If   A X = B   represents a system of equations where  A ≠ 0  and r = n

   then X = $A^{-1}$ B is an Unique Solution.

   A.   1. A X = B

        2. A ≠ 0 so $A^{-1}$  exists and is unique

        3. $A^{-1}$ [A X] = $A^{-1}$ [B]

        4. [$A^{-1}$ A] X = $A^{-1}$ B

        5. I X = $A^{-1}$ B

        6. X = $A^{-1}$ B

   B.   1. A is non-singular iff r = n

        2. also from previous a system has an Unique solution iff r = n

   C. therefore:  X = $A^{-1}$ B is an Unique Solution.

9.14. Adjoint of a Square Matrix $A_{nn}$  --- symbolized by  "Adj A"

I.

1. Write:
$$A = \begin{pmatrix} a_{11} & a_{12} & a_{13} & \ldots & a_{1n} \\ a_{21} & a_{22} & a_{23} & \ldots & a_{2n} \\ \ldots & \ldots & \ldots & \ldots \\ \ldots & \ldots & \ldots & \ldots \\ a_{i1} & a_{i2} & a_{i3} & \ldots & a_{in} \\ \ldots & \ldots & \ldots & \ldots \\ a_{k1} & a_{k2} & a_{k3} & \ldots & a_{kn} \\ \ldots & \ldots & \ldots & \ldots \\ \ldots & \ldots & \ldots & \ldots \\ a_{n1} & a_{n2} & a_{n3} & \ldots & a_{nn} \end{pmatrix}$$

where i can be any of 1,2,...,n

where k can be any of 1,2,...,n

2. Definition. The Adjoint of A where C is a Cofactor of A:

$$\text{Adj } A = \begin{pmatrix} C_{11} & C_{21} & C_{31} & \ldots & C_{n1} \\ C_{12} & C_{22} & C_{32} & \ldots & C_{n2} \\ C_{13} & C_{23} & C_{33} & \ldots & C_{n3} \\ \ldots & \ldots & \ldots & \ldots \\ C_{1j} & C_{2j} & C_{3j} & \ldots & C_{nj} \\ \ldots & \ldots & \ldots & \ldots \\ C_{1n} & C_{2n} & C_{3n} & \ldots & C_{nn} \end{pmatrix}$$

Note: we get this matrix by finding $A^T$ and
then replacing every $a_{ij}$ with its
Cofactor $C_{ij}$.

3. get a matrix B by replacing the kth row of A with another, say ith, row :

$$B = \begin{pmatrix} a_{11} & a_{12} & a_{13} & \cdots & a_{1n} \\ a_{21} & a_{22} & a_{23} & \cdots & a_{2n} \\ \cdots\cdots\cdots\cdots\cdots\cdots\cdots \\ \cdots\cdots\cdots\cdots\cdots\cdots\cdots \\ a_{i1} & a_{i2} & a_{i3} & \cdots & a_{in} \\ \cdots\cdots\cdots\cdots\cdots\cdots\cdots \\ a_{i1} & a_{i2} & a_{i3} & \cdots & a_{in} \\ \cdots\cdots\cdots\cdots\cdots\cdots\cdots \\ \cdots\cdots\cdots\cdots\cdots\cdots\cdots \\ a_{n1} & a_{n2} & a_{n3} & \cdots & a_{nn} \end{pmatrix}$$

4. because rwo rows are the same  $|B| = 0$,

5. evaluate $|B|$ by the expansion of cofactors of this new kth row; now  the elements of this kth row are now $a_{i1}$     , $a_{i2}$     ,...,$a_{in}$    while the cofactors of this kth row are still the [ $C_{k1}$     ,$C_{k2}$     ,... , $C_{kn}$ ] of A; so

$|B| = a_{i1} \ C_{k1} + a_{i2} \ C_{k2} + \ldots + a_{in} \ C_{kn} \ = 0$  where i & k are different numbers

5. note: if i = k : then this [ $a_{i1} \ C_{k1} + a_{i2} \ C_{k2} + \ldots + a_{in} \ C_{kn}$ ] is actually  $|A|$

II.

1. Examine:

$$[A][adj\ A] = \begin{pmatrix} a_{11} & a_{12} & a_{13} & \cdots & a_{1n} \\ a_{21} & a_{22} & a_{23} & \cdots & a_{2n} \\ \cdots\cdots\cdots\cdots\cdots\cdots\cdots \\ \cdots\cdots\cdots\cdots\cdots\cdots\cdots \\ a_{i1} & a_{i2} & a_{i3} & \cdots & a_{in} \\ \cdots\cdots\cdots\cdots\cdots\cdots\cdots \\ a_{k1} & a_{k2} & a_{k3} & \cdots & a_{kn} \\ \cdots\cdots\cdots\cdots\cdots\cdots\cdots \\ a_{n1} & a_{n2} & a_{n3} & \cdots & a_{nn} \end{pmatrix} \begin{pmatrix} C_{11} & C_{21} & C_{31} & \cdots & C_{n1} \\ C_{12} & C_{22} & C_{32} & \cdots & C_{n2} \\ \cdots\cdots\cdots\cdots\cdots\cdots\cdots \\ C_{1j} & C_{2j} & C_{3j} & \cdots & C_{nj} \\ \cdots\cdots\cdots\cdots\cdots\cdots\cdots \\ C_{1n} & C_{2n} & C_{3n} & \cdots & C_{nn} \end{pmatrix}$$

$$= \begin{pmatrix} d_{11} & \cdots & d_{1n} \\ \cdots\cdots\cdots\cdots \\ \cdots\cdots\cdots\cdots \\ d_{n1} & \cdots & d_{n1} \end{pmatrix}$$

2. note some of the d's of this product:

$$d_{11} = [ a_{11} \ C_{11} + a_{12} \ C_{12} + \ldots + a_{1n} \ C_{1n} ]$$

$$d_{12} = [ a_{11} \ C_{21} + a_{12} \ C_{22} + \ldots + a_{1n} \ C_{2n} ]$$

$$d_{21} = [ a_{21} \ C_{11} + a_{22} \ C_{12} + \ldots + a_{2n} \ C_{1n} ]$$

$$d_{22} = [ a_{21} \ C_{21} + a_{22} \ C_{22} + \ldots + a_{2n} \ C_{2n} ]$$

................................................

................................................

and thus:

$$d_{ij} = [ a_{i1} \ C_{j1} + a_{i2} \ C_{j2} + \ldots + a_{in} \ C_{jn} ] = \ |A| \quad if \quad i = j$$

$$= \quad 0 \quad if \quad i \neq j$$

so   $d_{11} = d_{22} = \ldots = d_{nn} = |A|$     because  $i = j$

while all the other elements of the product = 0  because  $i \neq j$

3. then after the product:

$$[A] \ [ \ adj \ A \ ] = \begin{pmatrix} |A| & 0 & 0 & \ldots & 0 \\ 0 & |A| & 0 & \ldots & 0 \\ 0 & 0 & |A| & \ldots & 0 \\ & \multicolumn{3}{c}{\ldots\ldots\ldots\ldots} & \\ & \multicolumn{3}{c}{\ldots\ldots\ldots\ldots} & \\ 0 & 0 & 0 & \ldots & |A| \end{pmatrix}$$

4. then divide by |A|:

$$A \cdot \{[1/|A|] \ [ \ adj \ A \ ]\} = \begin{pmatrix} 1 & 0 & 0 & \ldots & 0 \\ 0 & 1 & 0 & \ldots & 0 \\ 0 & 0 & 1 & \ldots & 0 \\ & \multicolumn{3}{c}{\ldots\ldots\ldots} & \\ & \multicolumn{3}{c}{\ldots\ldots\ldots} & \\ 0 & 0 & 0 & \ldots & 1 \end{pmatrix} = \ I_n \quad where \ |A| \neq 0$$

5. from the definition of an Inverse:

$$A^{-1} = [1/A] \text{ adj } A \quad : \quad |A| \neq 0$$

III. Example: find $A^{-1}$ given

$$A = \begin{bmatrix} 5 & -1 & 3 \\ 1 & -1 & 1 \\ 15 & -4 & 18 \end{bmatrix}$$

1. $C_{11} = + \begin{bmatrix} -1 & 1 \\ -4 & 18 \end{bmatrix} = -18 + 4 = -14$    $C_{12} = - \begin{bmatrix} 1 & 1 \\ 15 & 18 \end{bmatrix} = -18 + 15 = -3$

2. verify:

$$C_{13} = 11$$

$$C_{21} = 6 \qquad C_{22} = 45 \qquad C_{23} = 5$$

$$C_{31} = 2 \qquad C_{32} = -2 \qquad C_{33} = -4$$

3. as in part of 9.6  verify:   $|A| = -34$

4. so:

$$A^{-1} = [ 1/-34 ] \begin{bmatrix} -14 & 6 & 2 \\ -3 & 45 & -2 \\ 11 & 5 & -4 \end{bmatrix}$$

5. and so

$$A^{-1} = \begin{bmatrix} 7/17 & -3/17 & -1/17 \\ 3/34 & -45/34 & 2/34 \\ -11/34 & -5/34 & 2/17 \end{bmatrix}$$

9.15 Of special interest is the development of $D^{-1}$ of the Diagonal Matrix D.

A. Given the Diagonal matrix

$$D = \begin{bmatrix} a_{11} & 0 & 0 & \cdots & 0 & 0 \\ 0 & a_{22} & 0 & \cdots & 0 & 0 \\ 0 & 0 & a_{33} & \cdots & 0 & 0 \\ \cdots & \cdots & \cdots & \cdots & \cdots & \cdots \\ 0 & 0 & 0 & \cdots & a_{n-1\ n-1} & 0 \\ 0 & 0 & 0 & \cdots & 0 & a_{n\ n} \end{bmatrix}$$

B. to evaluate $D^{-1}$ we need the $C_{11}$ and $|D|$; find the $C_{11}$ first:

I. a. 1.

$$C_{11} = [-1]^{1+1} \begin{vmatrix} a_{22} & 0 & \cdots & 0 \\ 0 & a_{33} & \cdots & 0 \\ \cdots & \cdots & \cdots & \cdots \\ 0 & 0 & \cdots & a_{n\ n} \end{vmatrix} \quad \text{by the definition of cofactors}$$

2. evaluate the determinant by the expansion by cofactors of the 1st row:

$$C_{11} = + [-1]^{2+2}\ a_{22} \begin{vmatrix} a_{33} & \cdots & 0 \\ \cdots & \cdots & \cdots \\ 0 & \cdots & a_{n\ n} \end{vmatrix} + 0 + 0 + \cdots + 0,$$

$$C_{11} = + [-1]^{3+3}\ a_{22}\ a_{33} \begin{vmatrix} a_{44} & \cdots & 0 \\ \cdots & \cdots & \cdots \\ 0 & \cdots & a_{n\ n} \end{vmatrix} + 0 + \cdots + 0,$$

$$C_{11} = + [-1]^{[n-1]+[n-1]}\ a_{22}\ a_{33}\ a_{44}\ \cdots a_{n-1\ n-1}\ |a_{n\ n}|$$

3. $C_{11} = a_{22}\ a_{33}\ a_{44}\ \cdots a_{n-1\ n-1}\ a_{n\ n}$ note this is all the a's except $a_{11}$

b. in the same manner we can show that:

$C_{22} = a_{11}\ a_{33}\ a_{44}\ \cdots a_{n-1\ n-1}\ a_{n\ n}$ note this is all the a's except $a_{22}$

c. It is clear that :

$C_{ii} = [\ a_{11}\ a_{22}\ a_{33}\ a_{44}\ \cdots a_{n-1\ n-1}\ a_{n\ n}\ ]\ /\ a_{ii}$ for i = 1, 2, 3, ..., n.

II. Consider $C_{ij}$ where $i \neq j$

a.

1.

$$C_{12} = [-1]^{1+2} \begin{vmatrix} 0 & 0 & 0 & \ldots & 0 \\ 0 & a_{33} & 0 & \ldots & 0 \\ 0 & 0 & a_{44} & \ldots & 0 \\ \ldots & \ldots & \ldots & \ldots & \ldots \\ 0 & 0 & 0 & \ldots & a_{nn} \end{vmatrix} = 0$$

The 1, 2 in $C_{12}$ means that the 1st row is missing so no $a_{11}$ , and the 2 column is missing so no $a_{22}$ ; thus the 2nd row is all 0's.

2.

$$C_{13} = [-1]^{1+3} \begin{vmatrix} 0 & a_{22} & 0 & \ldots & 0 \\ 0 & 0 & 0 & \ldots & 0 \\ 0 & 0 & a_{44} & \ldots & 0 \\ \ldots & \ldots & \ldots & \ldots & \ldots \\ 0 & 0 & 0 & \ldots & a_{nn} \end{vmatrix} = 0$$

3.

$$C_{21} = [-1]^{2+1} \begin{vmatrix} 0 & 0 & 0 & \ldots & 0 \\ 0 & a_{33} & 0 & \ldots & 0 \\ 0 & 0 & a_{44} & \ldots & 0 \\ \ldots & \ldots & \ldots & \ldots & \ldots \\ 0 & 0 & 0 & \ldots & a_{nn} \end{vmatrix} = 0$$

4.

$$C_{32} = [-1]^{3+2} \begin{vmatrix} a_{11} & 0 & 0 & \ldots & 0 \\ 0 & 0 & 0 & \ldots & 0 \\ 0 & 0 & a_{44} & \ldots & 0 \\ \ldots & \ldots & \ldots & \ldots & \ldots \\ 0 & 0 & 0 & \ldots & a_{nn} \end{vmatrix} = 0$$

b. we see that for $C_{ij}$ , $i \neq j$ , $a_{11}$ is always eliminated so that the j row is always all 0's ; so $C_{ij} = 0$ : $i \neq j$ for $i$ , $j = 1, 2, \ldots, n$

C. Special Symbols.

1. The Summation or Sigma symbol: we have seen that

$$\sum_{1}^{n} a_k = a_1 + a_2 + \ldots + a_n .$$

2. The Product symbol:

$$\prod_{1}^{n} a_k = a_1 \cdot a_2 \cdot a_3 \cdots a_n .$$

3. These are "shorthand" symbols and do shorten much of the writing of the material in linear algebra. It has been felt that for beginning students of this kind of material that the expanded summation form is easier to understand --- rather than the single symbol representing a summation concept. However, for $D^{-1}$ the product symbol is very usefully; without it the evaluation becomes very messy.

D.

1. Use the 1$^{st}$ row of section 9.12.A.    $|D| = [-1]^{1+1} a_{11} C_{11} + 0 + 0 + \ldots + 0$

$$= + [-1]^{1+1} a_{11} [a_{22} \quad a_{33} \quad \ldots \quad a_{nn}]$$

or    $|D| = a_{11} \, a_{22} \, a_{33} \, \ldots \, a_{nn} = \prod_{1}^{n} a_{kk}$

2. now we can also write:

$$C_{ij} = \begin{cases} \prod_{1}^{n} a_{kk} / a_{ij} & : i = j \\ \\ 0 & : i \neq j \end{cases}$$

3. thus    if  $d_{ij}{}'$   represents the elements of $D^{-1}$  , then

   from 9.11. II. 5.:    $d_{ij}{}' = 0 : i \neq j$     and

$$d_{ii}{}' = \frac{C_{ii}}{|D|} = \frac{1}{a_{ii}} \quad : i = 1,2, \ldots, n$$

E.

$$D^{-1} = \begin{bmatrix} 1/a_{11} & 0 & \ldots & 0 \\ 0 & 1/a_{22} & \ldots & 0 \\ \ldots & \ldots & \ldots & \ldots & \ldots & . \\ \ldots & \ldots & \ldots & \ldots & \ldots & . \\ 0 & 0 & \ldots & 1/a_{nn} \end{bmatrix}$$

9.16 Two theorems.

A. Show $B = kA$ implies $B^{-1} = A^{-1}/k$ : $k \neq 0$

   1. $B = kA$

   2. $B^{-1} B = B^{-1} A k$

   3. $I = k B^{-1} A$

   4. $I A^{-1} = k B^{-1} A A^{-1}$

   5. $A^{-1} = k B^{-1} I$

   6. $A^{-1} = k B^{-1}$

   7. $[1/k] A^{-1} = B^{-1}$ : $k \neq 0$


B. Show that $[A^2]^{-1} = [A^{-1}]^2$

  I. $[A^2]^{-1} = [A \cdot A]^{-1} = A^{-1} \cdot A^{-1} = [A^{-1}]^2$

  II. show true if $A = \begin{bmatrix} 2 & 1 \\ 5 & 3 \end{bmatrix}$

   1. $A^2 = \begin{bmatrix} 2 & 1 \\ 5 & 3 \end{bmatrix} \begin{bmatrix} 2 & 1 \\ 5 & 3 \end{bmatrix} = \begin{bmatrix} 9 & 5 \\ 25 & 14 \end{bmatrix}$

   2. $|A^2| = 9 \cdot 14 - 5 \cdot 25 = 126 - 125 = 1$

   3. $[A^2]^{-1} = [1/1] \begin{bmatrix} 14 & -5 \\ -25 & 9 \end{bmatrix} = \begin{bmatrix} 14 & -5 \\ -25 & 9 \end{bmatrix}$

   4. $|A|$ $6 - 5 = 1$ so $A^{-1} = [1/1] \begin{bmatrix} 3 & -1 \\ -5 & 2 \end{bmatrix} = \begin{bmatrix} 3 & -1 \\ -5 & 2 \end{bmatrix}$

   5. $[A^{-1}]^2 = \begin{bmatrix} 3 & -1 \\ -5 & 2 \end{bmatrix} \begin{bmatrix} 3 & -1 \\ -5 & 2 \end{bmatrix} = \begin{bmatrix} 14 & -5 \\ -25 & 9 \end{bmatrix}$

   6. Verified that : $[A^2]^{-1} = [A^{-1}]^2$


9.17. Solve the system $\begin{cases} 2x + 4y + z = -2 \\ x + 2y + z = 2 \\ 3x + 4y + 2z = -4 \end{cases}$ by using the inverse concept.

  1. the system: $\begin{bmatrix} 2 & 4 & 1 \\ 1 & 2 & 1 \\ 3 & 4 & 2 \end{bmatrix} \begin{bmatrix} x \\ y \\ z \end{bmatrix} = \begin{bmatrix} -2 \\ 2 \\ -4 \end{bmatrix}$

let:
$$A = \begin{bmatrix} 2 & 4 & 1 \\ 1 & 2 & 1 \\ 3 & 4 & 2 \end{bmatrix} \quad ; U = \begin{bmatrix} x \\ y \\ z \end{bmatrix} \quad ; K = \begin{bmatrix} -2 \\ 2 \\ -4 \end{bmatrix}$$

then:  $A\ U\ =\ K$          in this system,

$$A^{-1}\ A\ U\ =\ A^{-1}\ K$$

$$I\quad U\ =\ A^{-1}\ K$$

$$U\ =\ A^{-1}\ K$$

2. $|A|\ =\ 8 + 12 + 4 - 6 - 8 - 8\ =\ 2$

3. $C_{11}\ =\ 0$ ,  $C_{21}\ =\ -4$ ,  $C_{31}\ =\ 2$

   $C_{12}\ =\ 1$ ,  $C_{22}\ =\ 1$ ,  $C_{32}\ =\ -1$

   $C_{13}\ =\ -2$ ,  $C_{23}\ =\ 4$ ,  $C_{33}\ =\ 0$

4.
$$A = [1/2] \begin{bmatrix} 0 & -4 & 2 \\ 1 & 1 & -1 \\ -2 & 4 & 0 \end{bmatrix} = \begin{bmatrix} 0 & -2 & 1 \\ ½ & ½ & -1/2 \\ -1 & 2 & 0 \end{bmatrix}$$

5.
$$\begin{bmatrix} x \\ y \\ z \end{bmatrix} = \begin{bmatrix} 0 & -2 & 1 \\ ½ & ½ & -1/2 \\ -1 & 2 & 0 \end{bmatrix} \begin{bmatrix} -2 \\ 2 \\ -4 \end{bmatrix}$$

6. $x = -8$ ,  $y = 2$ ,  $z = 6$

9.18 Loose Ends.

I. Scalar Product.

1. Given  $|A| = \begin{vmatrix} x & y \\ z & w \end{vmatrix} = xw - yz$

   we note that:  $\begin{vmatrix} kx & ky \\ z & w \end{vmatrix} = kxw - kyz = k\,[xw - yz] = k \begin{vmatrix} x & y \\ z & w \end{vmatrix} = k\,|A|$

2. It is clear that if the k factor had been on the $2^{nd}$ row instead the k |A| result would have been the same; but, if it had been on both rows, the result would have been   $k^2$  |A|.

   3. In general.  given  |A| =  |$a_{ij}$| for nxn     then

$$= \; + \; a_{11} \; C_{11} \; - \; a_{12} \; C_{12} \; + \; ... \; + \; [-1]^{1+n} \quad a_{1n} \; C_{1n}$$

   a. so:               $k \; a_{11} \; C_{11} \;\; - \; k \; a_{12} \; C_{12} \;\; + \; ... \; + \; [-1]^{1+n} \quad k \; a_{1n} \; C_{1n}$

   b. can be written:    $k \; ( \; a_{11} \; C_{11} \; - \; a_{11} \; C_{11} \; + \; ... \; + \; [-1]^{1+n} \quad a_{11} \;\; C_{11} \; )$

   c. thus equals        $k$ |A|


we see that the product between a Scalar and a Determinant is the product of every element of a row of your choice and the scalar.

   4. We have seen that with matrices every element of every row has to be multiplied by the scalar.

II.

   1. As also different from matrices, it is not possible to define a sum of determinants in terms of the determinants. We have to expand the determinant to a number and then use the algebra of real numbers.

   2. Without addition of |A| & |B| in terms of |A| & |B| we can not test for an abstract vector space nor for linearity in determinants.

10.1 INTRODUCTION.

I.

1. From the first via nomenclature and text we have indicated that the study of functions that satisfy the very important Linear Property would be our prime goal. However, we needed more than the real number system to act as our domains of these functions; so we studied arrows and matrices as sets of "objects" that also could be used.

2. In our quest for a general theory of Linear Functions we will borrow from general function theory what we need. We need to review and extend some of the concept of functions as used before. Here we will only be interested in functions that satisfy the Linear Property[or Identity]. Where possible we will use parenthesis f(x) for functions, [curly] braces { } for symbols of grouping, and brackets [ ] for matrices. We use xy, {}•{}, and [][] for Products and (x,y) for points with <x,y> for vectors.

3. When dealing with Linearity ,it is common practice to use the "transformation" nomenclature for functions and T for the symbol. We will do that here. Review 1.1 - 1.9.

4. All of the Spaces to be used as Domains in this text are assumed to be Abstract Vector Spaces without so stating. Beyond the value of being able to use the basic theorems of such a space, we can use either a vector or a row [ or column ] matrix for our pre-image and skip back and forth between the two in our work.

5. In this text for convenience we have restricted our number system to that of the Reals even though the concepts are true also for Complex numbers; so covered in Book II.

II. Some general symbols and definitions as we will use them in this text:

1. Given: X , Y , W ,... as elements of an Abstract Vector Space **S** when used as the independent variables of a Transformation rule T [or function f]. If the elements are matrices, they are of the form $X = [ x_{ij} ]_{m \times 1}$ , $Y = [ y_{ij} ]_{m \times 1}$ , $W = [ w_{ij} ]_{m \times 1}$ , ... If they are vectors, they are of the form $X = <x_1 , x_2 , ..., x_n>$ , $Y = < y_1 , y_2 , ..., y_n >$ , ... In all cases the Lower Case Letters represent Scalars [as in $X = < -1, 2, 5>$ ; so $x_1$ represents the first component of vector X, etc]. If Y is a subset of X, it includes of course the possibility that Y = X .

2. T(X) = Y can be used to symbolize the linear relation that for every X the rule T transforms each X into an unique { one & only one } Y.

3. The set of all X's that can be used is called the Domain or Primitive Space of the transformation T, and each X is the Independent Variable or Pre-Image of T. The set of

all Y's is called the Range or Image Space of T, and each Y is the Dependent Variable or

Image of its predecessor element X in the transformation T. The entire set of Y's is the

Codomain, and the Range is a subset of this set.

4. When each X has a different Y, we say that the transformation is One-to-One. If the

range is the entire set of Y's , we say that T is a transformation of X Onto Y . If the

range is a Proper Subset of all the Y's , the T is a transformation of X Into Y – that is

there is at least one of the Y's that is not in the range of T. In general in this text

we are not interested in such distinctions.

5. The use of the word Mapping to describe our function indicates that we are also

interested in the geometric nature of our transformation. The independent variable of our

function could be an ordered pair of numbers, and the transformation maps the points of

one plane onto points of another plane. Or, the mapping could be the transformation of

points in a plane to other points in the same plane.

III. Special Definitions for Functions in general.

1. Equality: Given: $T_1 (X) = Y$ & $T_2 (X) = Y$ , then $T_1 = T_2$

             iff    $T_1 (x_i ) = T_2 (x_i )$    for each given $x_i$ of X.

 2. The Zero Transformation is the T where $T (X) = 0$ for all X.

 3. The Identity Transformation is the T where $T (X) = X$.

10.2 Linear Transformation.

 I. Definition. Given: the transformation T to operate on The Abstract Vector

Space S whose elements are: X , X' , X'' , ...

          Then : this T is a Linear Transformation

          iff   $T ( X + X' ) = T ( X ) + T ( X' )$ :       the Additive Property

We Proved in Chapter One that if "the Additive Property" is true of a function, then

"the Homogeneuos Property" is also true. So,

     If  $T(X + X') = T(X) + T(X')$ ,the Additive Property of Linear Functions,

     Then $T(k X) = k T(X)$ , the Homogeneous Property of Linear Functions; k any scalar.

 Many times it is convenient to use both forms, but we only have to prove the Additive

Property to get Linearity. Without a proof most authors complicate this definition by

imposing on it the two conditions; this is not an acceptable thing for Definitions.

II. As we have seen before , a single, combined form of this Linear Transformation T

     would be :              $T (c X + k X' ) = c T (X ) + k T (X' )$.

III.  We will consider only single-variable functions; so while the vectors may be multi-
dimensional our functions involve only one independent vector and then from the rule one
resulting dependent vector. The Rule that defines the function will be a vector where
each component will be some linear combination of the components of the independent
vector--- some component(s) might be Zero. The number of components vary: 1,..., n.


10.3. Example:  Test: $T( <x, y, z> ) = <y, x-z, x>$

    1. $T( <x, y, z> + <x', y', z'> ) = T( <x + x', y + y', z + z'> )$

                                        $= <y + y', x + x' - z - z', x + x'>$

                                        $= <y, x - z, x> + <y', x' - z', x'>$

                                        $= T( <x, y, z> ) + T(<x', y', z'>)$

    2. so,  this T is Linear.

    3. Note that if the third component of T was "x + 1" and not "x", this T would not
be Linear. Check this out. Adding on a scalar in any component would give us a non-
linear transformation.


10.4 Theorem. A Linear Combination as a function is a Linear Transformation.

      Given:    $T(X) = a x_1 + b x_2 + ... + k x_n$

    Then: $T(X + X') = a \{x_1 + x_1'\} + b \{x_2 + x_2'\} + ... + k \{x_n + x_n'\}$

                      $= a x_1 + a x_1' + b x_2 + b x_2' + ... + k x_n + k x_n'$

                      $= \{a x_1 + b x_2 +...+ k x_n\} + \{a x_1' + b x_2' +...+ k x_n'\}$

                      $= T(X) + T(X')$     ;  so Linear.

 10.5 The power of having a linear transformation in function work increases the value of
the vector concept and the matrix concept when using these concepts to solve the problems
of the real world.


10.6 Linear Transformations as a Matrix Product.

 I.

    1. In 7.21 we proved that matrix multiplication is linear,

    2. so, If a transformation T is a Matrix product as in

      $T(X) = A X$   where $A = [a_{ij}]_{nxn}$     and $X = [x_{ij}]_{nx1}$ , then T is Linear.

II.   Given a Linear Transformation $T(X) = f(X)$ where $X = [x_{ij}]_{nx1}$ and f a given function   it

is possible to find a square matrix   $A = [a_{ij}]_{nxn}$   such that   $T(X) = A X$  . The matrix A   is

called The Matrix of the Transformation. It is clear that this is an equivalent/equal

relation; so  the equation   $T(X) = f(X)$ can always be replaced by the equation $A X = f(X)$.

   A.   A  Constructive Proof in 2-Space .

      1. Proof:

      a. Given $T(X)$ is a Linear Transformation; so from $T(X) = f(X)$ the function f also is

Linear. Thus, $f(X) = f \begin{bmatrix} x_1 \\ x_2 \end{bmatrix} = \begin{bmatrix} \{k_{11} x_1 + k_{12} x_2\} \\ \{k_{21} x_1 + k_{22} x_2\} \end{bmatrix}$   where the k's are given scalars

[includes k's = 0  ] for each component of the $f(X)$ vector has to be a Linear Combination of

the components of X ; thus these forms represent all possible linear functions in 2-space.

      b. so :   $T(X) = f(X)$ implies   $T \begin{bmatrix} x_1 \\ x_2 \end{bmatrix} = \begin{bmatrix} \{k_{11} x_1 + k_{12} x_2\} \\ \{k_{21} x_1 + k_{22} x_2\} \end{bmatrix}$   the k's are

                                                                          determined by T

      c. now $A X = \begin{bmatrix} a_{11} & a_{12} \\ a_{21} & a_{22} \end{bmatrix} \begin{bmatrix} x_1 \\ x_2 \end{bmatrix} = \begin{bmatrix} \{a_{11} x_1 + a_{12} x_2\} \\ \{a_{21} x_1 + a_{22} x_2\} \end{bmatrix}$

      d. we see we can easily construct the matrix A by letting $a_{11} = k_{11}$ , $a_{12} = k_{12}$ ,

         $a_{21} = k_{21}$ , & $a_{22} = k_{22}$ . There is no need to memorize this ; we  can find the

values of $a_{11}$ , $a_{12}$ , $a_{21}$ , & $a_{22}$ by inspection of the above method of lines a. & b.

      e. therefore $A X = T(X)$

      f. this method not only shows that this substitution can be done , but it shows how

to find the matrix A.

      g. this method can easily be extended to n-space by induction to complete the proof.

      2. Example: Given $T(X) = T \begin{bmatrix} x_1 \\ x_2 \end{bmatrix} = \begin{bmatrix} \{x_1 + 2x_2\} \\ \{-x_1\} \end{bmatrix}$ , Find , A , the Matrix of T.

         a. $A X = \begin{bmatrix} a & b \\ c & d \end{bmatrix} \begin{bmatrix} x_1 \\ x_2 \end{bmatrix} = \begin{bmatrix} \{a x_1 + b x_2\} \\ \{c x_1 + d x_2\} \end{bmatrix}$

         b. to make $A X = T(X)$ let a = 1 , b = 2  , c = -1  , d = 0  ;

         c. so $A = \begin{bmatrix} 1 & 2 \\ -1 & 0 \end{bmatrix}$

d. check: $AX = \begin{bmatrix} 1 & 2 \\ -1 & 0 \end{bmatrix} \begin{bmatrix} x_1 \\ x_2 \end{bmatrix} = \begin{bmatrix} x_1 + 2x_2 \\ -x_1 \end{bmatrix} = T \begin{bmatrix} x_1 \\ x_2 \end{bmatrix} = T(X)$

B.  A 2-Space Proof using a Basis that spans the space of the matrix X  [while not necessary as a proof, this method uses some devices that are well worth remembering].

1. Given T(X) a Linear Transformation; so $T(X + X') = T(X) + T(X')$ and $T(kX) = kT(X)$

2. let $X = \begin{bmatrix} x_1 \\ x_2 \end{bmatrix} = \begin{bmatrix} x_1 \cdot 1 + x_2 \cdot 0 \\ x_1 \cdot 0 + x_2 \cdot 1 \end{bmatrix} = \begin{bmatrix} x_1 \cdot 1 \\ x_1 \cdot 0 \end{bmatrix} + \begin{bmatrix} x_2 \cdot 0 \\ x_2 \cdot 1 \end{bmatrix} = x_1 \begin{bmatrix} 1 \\ 0 \end{bmatrix} + x_2 \begin{bmatrix} 0 \\ 1 \end{bmatrix}$

because the set $\left\{ \begin{bmatrix} 1 \\ 0 \end{bmatrix} , \begin{bmatrix} 0 \\ 1 \end{bmatrix} \right\}$ is a Basis for this  2-space.

3. thus $T(X) = T \left\{ x_1 \begin{bmatrix} 1 \\ 0 \end{bmatrix} + x_2 \begin{bmatrix} 0 \\ 1 \end{bmatrix} \right\} = T \left\{ x_1 \begin{bmatrix} 1 \\ 0 \end{bmatrix} \right\} + T \left\{ x_2 \begin{bmatrix} 0 \\ 1 \end{bmatrix} \right\}$ by #1a

$= x_1 T \left\{ \begin{bmatrix} 1 \\ 0 \end{bmatrix} \right\} + x_2 T \left\{ \begin{bmatrix} 0 \\ 1 \end{bmatrix} \right\}$ by #1b

4.  choose $[a_{ij}]_{n \times n}$ to be the labels assigned to the values of the following

transformations: $T \left\{ \begin{bmatrix} 1 \\ 0 \end{bmatrix} \right\} = \begin{bmatrix} a_{11} \\ a_{21} \end{bmatrix}$ and $T \left\{ \begin{bmatrix} 0 \\ 1 \end{bmatrix} \right\} = \begin{bmatrix} a_{12} \\ a_{22} \end{bmatrix}$

and  build a Matrix $A = (a_{ij})_{n \times n}$  so that $A = \begin{bmatrix} a_{11} & a_{12} \\ a_{21} & a_{22} \end{bmatrix}$

5. now $AX = \begin{bmatrix} a_{11} & a_{12} \\ a_{21} & a_{22} \end{bmatrix} \begin{bmatrix} x_1 \\ x_2 \end{bmatrix} = \begin{bmatrix} a_{11}x_1 + a_{12}x_2 \\ a_{21}x_1 + a_{22}x_2 \end{bmatrix} = \begin{bmatrix} a_{11}x_1 \\ a_{21}x_1 \end{bmatrix} + \begin{bmatrix} a_{11}x_2 \\ a_{22}x_2 \end{bmatrix}$

$= x_1 \begin{bmatrix} a_{11} \\ a_{21} \end{bmatrix} + x_2 \begin{bmatrix} a_{12} \\ a_{22} \end{bmatrix} = x_1 T \left\{ \begin{bmatrix} 1 \\ 0 \end{bmatrix} \right\} + x_2 T \left\{ \begin{bmatrix} 0 \\ 1 \end{bmatrix} \right\}$ by #4

6. from # 3 & # 5: $T(X) = AX$.  Note: in this material T{ } is serving as T( ) .

C. Uniqueness--- Is this matirix A unique?

   1. Assume there is another matrix $B = [\ b_{ij}\ ]_{nxn}$  such that $T(X) = B\ X$,

   2. then $B\ X = \begin{bmatrix} b_{11} & b_{12} \\ b_{21} & b_{22} \end{bmatrix} \begin{bmatrix} x_1 \\ x_2 \end{bmatrix} = \begin{bmatrix} b_{11}\ x_1 + b_{12}\ x_1 \\ b_{21}\ x_1 + b_{22}\ x_2 \end{bmatrix} = \begin{bmatrix} b_{11}\ x_1 \\ b_{21}\ x_1 \end{bmatrix} + \begin{bmatrix} b_{12}\ x_2 \\ b_{22}\ x_2 \end{bmatrix}$

$= x_1 \begin{bmatrix} b_{11} \\ b_{21} \end{bmatrix} + x_2 \begin{bmatrix} b_{12} \\ b_{22} \end{bmatrix} = x_1 \left\{ \begin{bmatrix} b_{11} & b_{12} \\ b_{21} & b_{22} \end{bmatrix} \begin{bmatrix} 1 \\ 0 \end{bmatrix} \right\} + x_2 \left\{ \begin{bmatrix} b_{11} & b_{12} \\ b_{21} & b_{22} \end{bmatrix} \begin{bmatrix} 0 \\ 1 \end{bmatrix} \right\}$

$B\ X = x_1\ B \begin{bmatrix} 1 \\ 0 \end{bmatrix} + x_2\ B \begin{bmatrix} 0 \\ 1 \end{bmatrix} = x_1\ T \left\{ \begin{bmatrix} 1 \\ 0 \end{bmatrix} \right\} + x_2\ T \left\{ \begin{bmatrix} 0 \\ 1 \end{bmatrix} \right\}$   from

the given

   3. from A.5:     $A\ X = x_1\ T \left\{ \begin{bmatrix} 1 \\ 0 \end{bmatrix} \right\} + x_2\ T \left\{ \begin{bmatrix} 0 \\ 1 \end{bmatrix} \right\}$

   4. so $B\ X\ =\ A\ X$ ;  therefore  A  is unique.

III. In summary:

   1. Given a Linear Transformation $T(X) = f(X)$ there exists a [square] Matrix,A, of T

where the product $A\ X\ =\ T(X)$. We find the specific A from the equation:

$$\begin{bmatrix} a_{11} & a_{12} & \ldots & a_{1n} \\ a_{21} & a_{22} & \ldots & a_{2n} \\ \ldots & \ldots & \ldots & \ldots \\ \ldots & \ldots & \ldots & \ldots \\ a_{n1} & \ldots & \ldots & a_{nn} \end{bmatrix} \begin{bmatrix} x_1 \\ x_2 \\ . \\ . \\ x_n \end{bmatrix} = f(X)$$

2. Example. Find the A of the T : $T(X)\ =\ \begin{bmatrix} -2x\ +\ x' \\ -\ x \end{bmatrix}$

a.  $A\ X = \begin{bmatrix} a & b \\ c & d \end{bmatrix} \begin{bmatrix} x \\ x' \end{bmatrix} = \begin{bmatrix} a\ x\ +\ b\ x' \\ c\ x\ +\ d\ x' \end{bmatrix} = \begin{bmatrix} -2x\ +\ x' \\ -\ x \end{bmatrix}$

   b. so  $a = -2$ ,  $b = 1$ ,  $c = -1$ ,  $d = 0$

   c. therefore  $A\ = \begin{bmatrix} -2 & 1 \\ -1 & 0 \end{bmatrix}$

   4. check: $A\ X = \begin{bmatrix} -2 & 1 \\ -1 & 0 \end{bmatrix} \begin{bmatrix} x \\ x' \end{bmatrix} = \begin{bmatrix} -2\ x\ +\ x' \\ -x \end{bmatrix} = T\ (\ X\ )$

IV.

A. Given the square matrix A it is possible to find a linear transformation T(X)

   such that A X = T(X).

B. the Method: given A $= \begin{bmatrix} 1 & 2 \\ -1 & 0 \end{bmatrix}$ find T(X) so that A X = T(X)

1.    A X $= \begin{bmatrix} 1 & 2 \\ -1 & 0 \end{bmatrix} \begin{bmatrix} x \\ x' \end{bmatrix} = \begin{bmatrix} x + 2x' \\ -x \end{bmatrix}$ because a matrix product it

is a linear transformation

2. therefore    T(X) $= \begin{bmatrix} x + 2x' \\ -x \end{bmatrix}$

10.7. Geometric Operations that are Linear Transformations.

I. Consider the Rotation of Cartesian Coordinate Axes in a Plane.

1. Given  a Coordinate System in a Plane with axes OX & OX' . Rotate this system of axes

about the origin O by the fixed angle ø to form the OY & OY'  axes. Take  P as any point in

the plane and list the coordinates as ( x ,x' ) in reference to the X,X' axes and

( y ,y' ) in reference to the Y,Y' axes;

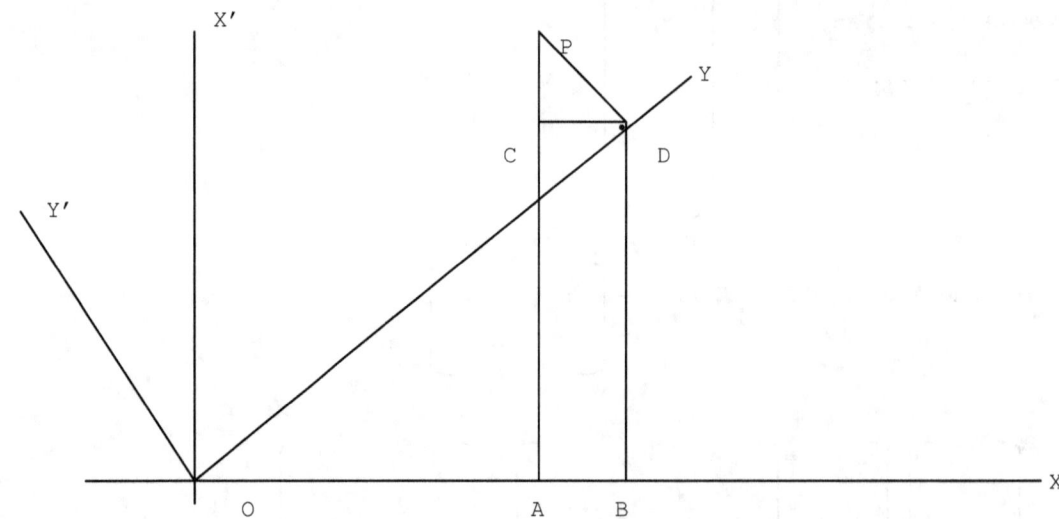

2. we see: the projection from the point P to the X axis the gives us the

points C & A ; the projection of P onto the OY axis gives us the point D ; the projection

of D onto the X axis gives point B; angle AOD is  ø  &  angle APD is also ø ;

3. we see that : OA = x , AP = x' , OD = y , DP = y' , AB = CD , AC = BD

4. In Analytic Geometry with the above figure we proved with many steps that

$$\left\{ \begin{array}{l} y \;\; = \; x \;\; \cos \; ø \; + \; x' \; \sin \; ø \\ y' \; = -x \;\; \sin \; ø \; + \; x' \; \cos \; ø \end{array} \right\}$$

5. To show this concept  by linear transformation forms we take

$$Y = \begin{bmatrix} y \\ y' \end{bmatrix} \;\; = \;\; T(X) \;\; = T \begin{bmatrix} x \\ x' \end{bmatrix} \;\; = \begin{bmatrix} \{ x \cos ø + x' \sin ø\} \\ \{-x \sin ø + x' \cos ø\} \end{bmatrix}$$

note: we can not use linear algebra to solve rotation, but rather to put the solution in

another form which might be of some better use.

6. and then build a matrix A by:

$$\begin{bmatrix} a & b \\ c & d \end{bmatrix} \begin{bmatrix} x \\ x' \end{bmatrix} = \begin{bmatrix} a x & + b x' \\ c x & + d x' \end{bmatrix}$$

7 . thus to get T(X) = A X we let a = cos ø , b = sin ø , c = - sin ø , d = cos ø

and get the specific A as:

$$A \;\; = \begin{bmatrix} \cos ø & \sin ø \\ -\sin ø & \cos ø \end{bmatrix}$$

8. we can write the new coordinates (y, y') as: Y = T(X) = A X,

9. we note that the Rotation is completely determined by the matrix A, because the

sines and the cosines are independent of the x's & y's,.    But, more importantly we

note that the Rotation of Axes is in a one-to-one correspondence with the matrix A by

means of the matrix equation Y  = A  X  , and so it is a Linear Transformation.

9. Example. With a rotation of axes of  30°   find the new coordinates of the point ( 2, 4 )

---   use 1.7 as the approximate value of   √3  :

a. now A   X  $= \begin{bmatrix} [\sqrt{3}]/2 & \frac{1}{2} \\ - \frac{1}{2} & [\sqrt{3}]/2 \end{bmatrix} \begin{bmatrix} 2 \\ 4 \end{bmatrix} = \begin{bmatrix} \{\sqrt{3}+2\} \\ \{-1+2\sqrt{3}\} \end{bmatrix} \approx \begin{bmatrix} 3.7 \\ 2.4 \end{bmatrix}$

b.   the (2, 4) point is now the  (3.7, 2.4) point. Note: the solution was a matrix

product instead of a subsition and then a simplification.

II. A Reflection problem  -- see the figure:

 1. Given this line L in the first quadrant of a

xy coordinate system; project each point of L

perpendicular to the x axis; extend each

projection an equal distance from the x axis;

these points define a new line L'  called the

Reflection of L through the x axis; points P'

are the Mirror Images of the points P through

the Mirror [the x axis].

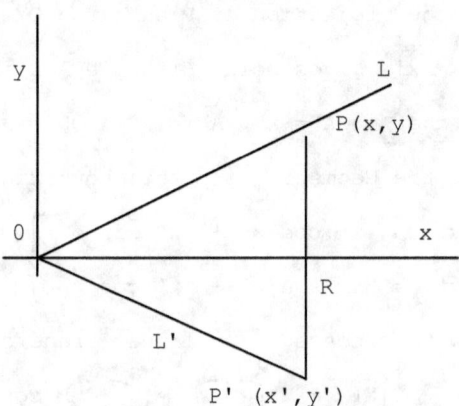

 2. we see that x' = x  & y' = - y; thus if (x,y) is represented by $\begin{bmatrix} x \\ y \end{bmatrix}$ then

the matrix equation $\begin{bmatrix} x' \\ y' \end{bmatrix} = \begin{bmatrix} 1 & 0 \\ 0 & -1 \end{bmatrix} \begin{bmatrix} x \\ y \end{bmatrix}$ represents the reflection.

3. Reflection is a Linear Transformation because of the matrix product.

4. If we know the equation of line L , then making the substitutions of step 2

we get the equation of line L'.

III. Projection - see same figure.

 1. From 2.16 the $Proj_{ox}$ OP = OR

 2.  we see that for the points (x'' , y'') of OR :    x'' = x & y'' = 0

 3. the matrix equation $\begin{bmatrix} x'' \\ y'' \end{bmatrix} = \begin{bmatrix} 1 & 0 \\ 0 & 0 \end{bmatrix} \begin{bmatrix} x \\ y \end{bmatrix}$ represents the projection.

 4. Projection  is a Linear Transformation.

IV.

     A. Many geometric problems can be handled in this manner, and thus are seen

to be linear transformations. In such work we would be interested in the various

A matrices that perform the transformations.

     B. Exercise. Find the matrix equation for the following:  Given the line L:

2x - y = 3 , Translate the Axes [ that is Move the Axes parallel to themselves ]

so that the Origin 0' of L' is at the point A(3,3).

     a. if x = 3 & y = 3 then in L: 6 - 3 = 3 or 3=3; so pt 0' is on the line L  --

this makes part of L' a vector in the Standard Position which we have found useful;

b. any point ( x',y') on the line will satisfy the equations x' = x - 3 & y' = y - 3

c. this system can be represented by the Matrix Equation $\begin{bmatrix} x' \\ y' \end{bmatrix} = \begin{bmatrix} x \\ y \end{bmatrix} + \begin{bmatrix} -3 \\ -3 \end{bmatrix}$

d. because the matrix equation is the sum of two matrices,

    Translation of Axes is not Linear.

10.8 Linear Transformations in terms of a Basis.

    A. We sometimes find the following is convenient in Theory Work when dealing with the Linear Function  Y = T(X) where as a vector $X = <x_1 , x_2 , x_3 , ..., x_n>$

B.

    1. we will use the Standard Orthonormal Basis for Column Vectors:

$$e_{(1)} = \begin{pmatrix} 1 \\ 0 \\ . \\ . \\ . \\ 0 \end{pmatrix} , \; e_{(2)} = \begin{pmatrix} 0 \\ 1 \\ . \\ . \\ . \\ 0 \end{pmatrix} , \; ... , \; e_{(n)} = \begin{pmatrix} 0 \\ 0 \\ . \\ . \\ . \\ 1 \end{pmatrix}$$

    2. note:  a. the "ortho" prefix refers to the fact that geometrically they are mutually Perpendicular to each other or algebraically any thing that means this perpendicular concept,

        b. the "Normal" suffix refers to the fact that they are all Unit Vectors.

C.   1.

        a. Now we can express any vector X  in terms of this Basis , as:

        $X = x_1 \; e_{(1)} + x_2 \; e_{(2)} + ... + x_n \; e_{(n)}$ where the $x_i$ are known scalars

    2. As we see:

$$X = x_1 \begin{pmatrix} 1 \\ 0 \\ . \\ 0 \end{pmatrix} + x_2 \begin{pmatrix} 0 \\ 1 \\ . \\ 0 \end{pmatrix} + ... + x_n \begin{pmatrix} 0 \\ 0 \\ . \\ 1 \end{pmatrix}$$

$$X = \begin{pmatrix} x_1 \\ 0 \\ . \\ 0 \end{pmatrix} + \begin{pmatrix} 0 \\ x_2 \\ . \\ 0 \end{pmatrix} + ... + \begin{pmatrix} 0 \\ 0 \\ . \\ x_n \end{pmatrix} = \begin{pmatrix} x_1 \\ x_2 \\ . \\ x_n \end{pmatrix}$$

    or   $X = < x_1, x_2, ... , x_n>$

2.

   a. If we need it we can do the same with Y. Note we could  use a different Basis
to resolve Y into components in the same problem, but it is easier to use the same.

   b. thus

$$Y = y_1 e_{(1)} + y_2 e_{(2)} + \ldots + y_n e_{(n)} = \begin{bmatrix} y_1 \\ y_2 \\ . \\ y_n \end{bmatrix}$$

D. From the Linear property:

$$Y = T(X) = T( x_1 e_{(1)} + x_2 e_{(2)} + \ldots + x_n e_{(n)} )$$

$$Y = T( x_1 e_{(1)} ) + T(x_2 e_{(2)} )+ \ldots + T(x_n e_{(n)} )$$

$$Y = x_1 T(e_{(1)} )+ x_2 T(e_{(2)} )+ \ldots + x_n T(e_{(n)} )$$

 In words to find the Image Y find the function F of the Basis Vectors e's ,and then
set up a Linear Combination of these terms with the components of the Pre-Image Vector X
in terms of that Basis.

E. Example. Find: $Y = T(<1,2>)$  if $T(<x_1 , x_2 >) = < x_1 + 2 x_2 , - x_1 >$

         Solution:

         1. could directly : $Y = < 1 + 2 \cdot 2, -1> = < 5 , -1 >$

         2. or using D. above: $Y = 1 T( <1, 0> ) + 2 T ( <0, 1> )$

$$= 1 \{ < 1, -1 > \} + 2 \{ < 2, 0 > \}$$

$$= < 1, -1 > + < 4, 0 > = < 5, -1 >$$

         3. In this problem it is easy to jump steps in #2 but still nothing is
gained by the second method; however if n is very large and T is complicated,

something might be gained because of the 1's  & the 0's in the e's.

10.9 The Algebra of Linear Transformations.

   A. Because every linear transformation T can be expressed as a Matrix product, we
can operate with linear transformations as we can with matrix products. In the process
we can build an algebra of linear transformations.

B. If we are working with the matrix product forms of linear transformations, then we need no new symbols nor theorems. If we desire to work with the T(X) forms, we need new new definitions/symbols. Because of the equivalent/equal matrix forms it seems reasonable to take the following symbolic forms to be true by Definition:

   1.    a. $\{T + T'\}$ (X) = T(X) + T'(X) ,

        b. $\{k\ T\}$ (X) = k $\{T(X)\}$ ,

        c. $\{T\ T'\}$ (X) = T $\{T'(X)\}$ .

  2. thus if T(X) = A X  and T'(X) = B X :

    a.  $\{T + T'\}$ (X) = T(X) + T'(X) = A X + B X = $\{A + B\}$ X,

    b. $\{k\ T\}$ (X) = k $\{T(X)\}$ = k $\{A\ X\}$ = $\{k\ A\}$ X,

    c. $\{T\ T'\}$ (X) = T $\{T'(X)\}$ = T $\{B\ X\}$ = $\{A\}$ $\{B\ X\}$ = $\{A\ B\}$ X.

  3. It is clear that if T & T' are linear transformations , then all of the expressions in #1 & #2 are also linear. Exercise: show that a possible defintion of

$$\{T\ T'\}\ (X)\ \longrightarrow\ \ T(X) \cdot T'(X)\ \ \text{for #2c  is Not Linear.}$$

  4. The end results in #2 are called  Composite Linear Transformations because of the sequence of operations performed.


10.10 Example [in form use rotation as a general composite linear transformation]:

    A.  1. take the conversion of a $w_1$  $w_1$  coordinate system into a $x_1$  $x_1$  system by:

$$x_1 = b_{11}\ w_1 + b_{12}\ w_2$$
$$x_2 = b_{21}\ w_1 + b_{22}\ w_2 \qquad ,$$

note: in a 2-space these equations represent the most general way

to change $w_1$ & $w_2$ in terms of x's no matter what kind of transformation is used,

    2. now take the conversion of the $x_1$ $x_2$  coordinate system into a $y_1$ $y_2$  system by:

$$y_1 = a_{11}\ x_1 + a_{12}\ x_2$$
$$y_2 = a_{21}\ x_1 + a_{22}\ x_2 \qquad ,$$

    3. then we can convert the $w_1$  $w_1$  coordinate system into the $y_1$  $y_1$  system  by:

      a. let $Y = \begin{bmatrix} y_1 \\ y_2 \end{bmatrix}$ ,  $X = \begin{bmatrix} x_1 \\ x_2 \end{bmatrix}$ ,  $W = \begin{bmatrix} w_1 \\ w_2 \end{bmatrix}$

        and make up an  $A = (a_{ij}\ )_{2 \times 2}$  & $B = (b_{jk}\ )_{2 \times 2}$

      b. then Y = A X  and X = B W ;

      c. thus   Y =  A $\{B\ W\}$ = $\{A\ B\}$ W

d. as in:

$$Y = \left[ \begin{bmatrix} a_{11} & a_{11} \\ a_{21} & a_{22} \end{bmatrix} \begin{bmatrix} b_{11} & b_{11} \\ b_{21} & b_{22} \end{bmatrix} \right] W$$

$$= \begin{bmatrix} (a_{11} b_{11} + a_{12} b_{21}) & (a_{11} b_{12} + a_{12} b_{22}) \\ (a_{21} b_{11} + a_{22} b_{21}) & (a_{21} b_{12} + a_{22} b_{22}) \end{bmatrix} \begin{bmatrix} w_1 \\ w_2 \end{bmatrix}$$

e. or $\begin{bmatrix} y_1 \\ y_2 \end{bmatrix} = \begin{bmatrix} (a_{11} b_{11} + a_{12} b_{21}) w_1 + (a_{11} b_{12} + a_{12} b_{22}) w_2 \\ (a_{21} b_{11} + a_{22} b_{21}) w_1 + (a_{21} b_{12} + a_{22} b_{22}) w_2 \end{bmatrix}$

f. so $\begin{bmatrix} y_1 \\ y_2 \end{bmatrix} = \begin{bmatrix} \{a_{11} b_{11} w_1 + a_{12} b_{21} w_1 + a_{11} b_{12} w_2 + a_{12} b_{22} w_2\} \\ \{a_{21} b_{11} w_1 + a_{22} b_{21} w_1 + a_{21} b_{12} w_2 + a_{22} b_{22} w_2\} \end{bmatrix}$

2. We could get this same Composite Linear Transformation if we substitute the x equation into the y equation by algebraic means , but in many cases it is easier to do the matrix muliplication.

3. A reminder: Y = A  X  can be written as  Y = T X to remind one that in this linear transformation the symbol T represents the operation of " multiply by A" and the symbol T X represents the result of such a multiplication.

4. So, prior to section 8.1 we saw how the matrix A could help us solve ,say, systems of equations after we learned to change their form and so on. Now the matrix also can be used to define the operation of multiplying by that matrix. That is why some use the symbol T to show this use and A to show the original use.

10.11 Examples.

A. Find the linear transformation that represents the following matrices and note the geometric interpretation of each.

1. $T = \begin{bmatrix} 0 & 1 \\ 1 & 0 \end{bmatrix}$                    see 11.7 #3 for the use of T and not A,

a. now $\begin{bmatrix} y_1 \\ y_2 \end{bmatrix} = \begin{bmatrix} 0 & 1 \\ 1 & 0 \end{bmatrix} \begin{bmatrix} x_1 \\ x_2 \end{bmatrix} = \begin{bmatrix} x_2 \\ x_1 \end{bmatrix}$ ;

b. so $T \begin{bmatrix} x_1 \\ x_2 \end{bmatrix} = \begin{bmatrix} x_2 \\ x_1 \end{bmatrix}$

c. T has changed $x_1$  into  $x_2$   and $x_2$  into  $x_1$

   d. draw a XOY coordinate system and show points P (x, y) & P' (x', y'). Take

vector OZ in the standard position with angle XOZ at 45°. Draw PP' cuts OZ at A. Draw OP

& OP'. Project P onto OX to get point B & P' onto OY to get point C. List angles:

BOP as α, POA as Φ, AOP' as δ, & P'OC as β. Angle AOC is also of 45°.

   e. $OP = \sqrt{x^2 + y^2}$ , $OP' = \sqrt{x^2 + y^2}$ ; so OP = OP'

   $\tan \alpha = y/x$ , $\tan \beta = y/x$ ; so α = β

   Φ = 45° - α, δ = 45° - β ; so Φ = δ

   OA ≡ OA , Δ POA congruent to ΔAOP'Φ ; so PP' _|_ OZ

   PA = AP' .

   f. so T is the reflection of points through the 45° angled  OZ

2. $T = \begin{bmatrix} 1 & 0 \\ 0 & -1 \end{bmatrix}$

   a. now $\begin{bmatrix} x' \\ y' \end{bmatrix} = \begin{bmatrix} 1 & 0 \\ 0 & -1 \end{bmatrix} \begin{bmatrix} x \\ y \end{bmatrix} = \begin{bmatrix} x \\ -y \end{bmatrix}$

   b.

   c. $T \begin{bmatrix} x' \\ y' \end{bmatrix} = \begin{bmatrix} x \\ -y \end{bmatrix}$

   d. T is the Reflection of points through the X-axis.

3. $T = \begin{bmatrix} a & 0 \\ 0 & 1 \end{bmatrix}$    a>0

   a. $\begin{bmatrix} y \\ y' \end{bmatrix} = \begin{bmatrix} a & 0 \\ 0 & 1 \end{bmatrix} \begin{bmatrix} x \\ x' \end{bmatrix} = \begin{bmatrix} a\,x \\ x' \end{bmatrix}$

   b. $T \begin{bmatrix} x \\ x' \end{bmatrix} = \begin{bmatrix} a\,x \\ x' \end{bmatrix}$

c. T describes the motion of a point parallel to the X-axis and if a >1 its final position is to the right of the original position. If  0< a <1, then the final position is between the x axis and the original position.. If a<0 then a similar action takes place to the left of the x-axis.

B. Find the matrix A that represents the Linear Transformation that maps the Point ( x , x' ) onto the point ( 2x  - 5x' , 3x  + 4x' ).

either $\begin{bmatrix} a & b \\ c & d \end{bmatrix}\begin{bmatrix} x \\ x' \end{bmatrix} = \begin{bmatrix} 2x - 5x' \\ 3x + 4x' \end{bmatrix}$ => $\begin{bmatrix} ax + bx' \\ cx + dx' \end{bmatrix} = \begin{bmatrix} 2x - 5x' \\ 3x + 4x' \end{bmatrix}$ => a = 2 ,b = -5    c =3  ,d = 4

; so  A = $\begin{bmatrix} 2 & -5 \\ 3 & 4 \end{bmatrix}$

or $\begin{cases} y = 2x - 5x' \\ y' = 3x + 4x' \end{cases}$ => $\begin{bmatrix} y \\ y' \end{bmatrix} = \begin{bmatrix} 2 & -5 \\ 3 & 4 \end{bmatrix}\begin{bmatrix} x \\ x' \end{bmatrix}$ ; so  A = $\begin{bmatrix} 2 & -5 \\ 3 & 4 \end{bmatrix}$

the geometry is too complicated to discuss.

C. Given the equations $\begin{cases} y = x - 2x' \\ y' = -x + x' \end{cases}$ and $\begin{cases} x = 3w - w' \\ x' = w - 2w' \end{cases}$

Find  y & y'  in terms of w & w'

Solution:

1. let Y = $\begin{bmatrix} y \\ y' \end{bmatrix} = \begin{bmatrix} 1 & -2 \\ -1 & 1 \end{bmatrix}\begin{bmatrix} x \\ x' \end{bmatrix}$ and  X = $\begin{bmatrix} x \\ x' \end{bmatrix} = \begin{bmatrix} 3 & -1 \\ 1 & -2 \end{bmatrix}\begin{bmatrix} w \\ w' \end{bmatrix}$

2. so Y = $\begin{bmatrix} 1 & -2 \\ -1 & 1 \end{bmatrix}\begin{bmatrix} 3 & -1 \\ 1 & -2 \end{bmatrix}\begin{bmatrix} w \\ w' \end{bmatrix} = \begin{bmatrix} [3-2] & [-1+4] \\ [-3+1] & [1-2] \end{bmatrix}\begin{bmatrix} w \\ w' \end{bmatrix}$

= $\begin{bmatrix} 1 & 3 \\ -2 & -1 \end{bmatrix}\begin{bmatrix} w \\ w' \end{bmatrix}$ ; thus $\begin{cases} y = w + 3w' \\ y' = -2w - w' \end{cases}$

10.12 Multi-variable functions.

I. Most of our work can be expanded to included Transformations that involved more than one independent variable in the domain of our function.

II. Theorem. Given the vector space V that maps onto the vector space U by the linear transformation T.

If    the set $X = \{ X_1, X_2, ..., X_n \}$ is a Basis of V

Then  the set $Y = \{ T(X_1), T(X_2), ..., T(X_n) \}$ Spans U.

Proof:

1. because of the definition of a Basis the set $X = \{ X_1, X_2, ..., X_n \}$ is Linear Independent,

2. any vector $A \; \varepsilon \; V$ is a Linear Combination of the Basis, or

   $A = a_1 X_1 + a_2 X_2 + ... a_n X_n$   where $A = 0$ if $a_1 = a_2 = ... = a_n = 0$

3. $T(A) = a_1 T(X_1) + a_2 T(X_2) + ... + a_2 T(X_2)$ because T is a Linear

Transformation; so T(A) is any vector in U,

4. the $T(X_i) : i = 1,2,...,n = 0$ ; so $T(A) = 0$ iff $a_1 = a_2 = ... = a_n = 0$

5. thus the set $\{ T(X_1), T(X_2), ..., T(X_n) \}$ is Linear Independent,

6. therefore the set Y Spans U by the definition of Span.

## 11.0 Introduction.

1. At the beginning of this text we indicated that a very important way to solve a great number of the problems of the real world was to first find a mathematical model of the problem. While not discussed at that time, the model we were referring to was usually a set of one or more conditional equations; or: expressions involving known and unknown quantities which describe an identity on the condition that the unknown(s) have certain value(s).

2. There are all sorts of classes of equations progressing from the very simple $y = kx$ to the quadratic to the differential to beyond what the student has studied so far. The search for methods of solution have lead mathematicians further to the search for general methods of solutions of equations. This can be difficult.

3. Fortunately a great number of equations to be solved from the real world can be handled by the use of the methods of Linear Algebra. Problems that are Linear in nature are basically Linear Transformations where we are considering equations as functions.

## 11.1 Eigenvalues/Eigenvectors.

I.   1. The simplest linear transformation is $T(X) = X : X \neq 0$. As we have seen before, it is possible to find a square matrix A so that the matirix product A X will accomplish the same thing as following the rule T on X to obtain the desired result: as A X = X. This function is defined completely by the rule T or multiplier matrix A that transforms the Independent Variable into Itself.

2. We see that $T(X)$ , A X , & X  are all the same vector; geometrically they are parallel. As we use the concept in this text X & T(X) & A X are equal vectors on the same line; a line that passes through the origin.  X, T(X), & A X are equivalent/equal.

3. In this case we say that X is Invariant under this linear transformation. No matter what we take for the  variable, X,   T(X) & A X give the same expression.

II.

1. We consider that the equation $T(X) = \lambda X$  [or $A X = \lambda X$] : A  square, $\lambda$  a  scalar, and $X \neq 0$ to be the general case of the equation in step #1. This kind of equation is what we are now interested in studying. In this case T(X), A X, and  X  are Invariants. What is constant is the line of the vector X that passes through the origin  and not X's length or direction. This A X and the X are Equivalent; this is not generally true.

   The geometry of our work in this chapter states the we have a set of vectors $\lambda$ A, all with different $\lambda$'s, and all are acting along the same straight line through the origin. Now this seems too special to be of use in solving the problems of the real world. However, as the student studies more and more mathematics, physics, engineering, ... , he will find a great number of real problems that fit this case. As an example consider the brakes of a car. Car designers would study all force vectors that act along the same straight line; the straight line travel of the car before it slows or stops.

   2. In our work each X will be a row vector, $\langle x_{ij}\rangle_{1 \times n}$ , or a column matrix, $(x_{ij})_{n \times 1}$ , while A is the square matrix $(a_{ij})_{n \times n}$   .

   3. One kind of problem we have here is: given a linear transformation $T(X) = \lambda X$ that is defined by a given matrix A in the matrix equation $A X = \lambda X$, find all particular combinations of $\lambda$ / X  that are  solutions of the matrix equation.

   4. This scalar $\lambda$  is very important in many circumstances and is called the Eigenvalue of T [ or of A ] with respect to a particular solution X.  Some writers prefer the nomenclature : $\lambda$  is the Characteristic Value of T  of X.

   5. The corresponding , non-zero solution, X , of either form of the equation is called the Eigenvector of T with respect to its eigenvalue $\lambda$. Characteristic Vector , Proper Vector, or Latent Root are terms that are also used. Note: the German word Eigen means many things; as: special, characteristic, particular , and the word Proper is used as in "one's own" or "distinctive".

   6. We will start our study of Eigenvalues/Eigenvectors by working in 2 & 3 space. It will be clear at all times how to expand these materials to n-space --- even though we will lose our geometry and the algebra may become difficult.

11.2 Illustrations & Examples.

 I. Examine the equation of the linear transformation:       $T \begin{bmatrix} x_1 \\ x_2 \end{bmatrix} = \begin{bmatrix} -x_1 \\ -x_2 \end{bmatrix}$

   Draw the geometry and note T is the Reflection of a point through the Origin.

   1. the corresponding the matrix product A X  is:   $\begin{bmatrix} a & b \\ c & d \end{bmatrix}\begin{bmatrix} x_1 \\ x_2 \end{bmatrix} = \begin{bmatrix} -x_1 \\ -x_2 \end{bmatrix}$

        and thus  $a x_1 + b x_2 = -x_1$   and  $c x_1 + d x_2 = -x_2$

using previous methods we see that a = -1 , b = 0 , c = 0 , & d = -1 ;

so the equation is also of the form: $\begin{bmatrix} -1 & 0 \\ 0 & -1 \end{bmatrix} \begin{bmatrix} x_1 \\ x_2 \end{bmatrix} = \begin{bmatrix} -x_1 \\ -x_2 \end{bmatrix} = -1 \begin{bmatrix} x_1 \\ x_2 \end{bmatrix}$

2. thus for the eigenvector X = $\begin{bmatrix} x_1 \\ x_1 \end{bmatrix}$ of this transformation T the Eigenvalue is -1.

3. now: $T \begin{bmatrix} \{x_1 + x_1'\} \\ \{x_1 + x_1'\} \end{bmatrix} = \begin{bmatrix} -\{x_1 + x_1'\} \\ -\{x_1 + x_1'\} \end{bmatrix} = \begin{bmatrix} \{-x_1 - x_1'\} \\ \{-x_1 - x_1'\} \end{bmatrix}$

$= \begin{bmatrix} \{-x_1\} + \{-x_1'\} \\ \{-x_1\} + \{-x_1'\} \end{bmatrix} = T \begin{bmatrix} x_1 \\ x_1 \end{bmatrix} + T \begin{bmatrix} x_1' \\ x_1' \end{bmatrix}$

4. thus T is Linear.

II. Example. Given a linear transformation T represented by A X = $\lambda$ X where A = $\begin{bmatrix} 3 & 0 \\ 0 & 3 \end{bmatrix}$

   A.   Show that one eigenvector is X = < 3 , 2 > ; find its eigenvalue $\lambda$ .

   1. A X = $\begin{bmatrix} 3 & 0 \\ 0 & 3 \end{bmatrix} \begin{bmatrix} 3 \\ 2 \end{bmatrix} = \begin{bmatrix} 9 \\ 6 \end{bmatrix} = 3 \begin{bmatrix} 3 \\ 2 \end{bmatrix} = 3$ X

   2. so for this T: X = < 3 , 2> is an eigenvector and its eigenvalue is 3

   3. geometrically this T stretches the vector <3 , 2 > to three times its

size along the same line through the origin and in the same direction.

   B. Test X = < 6 , 4 >.

   1. A X = $\begin{bmatrix} 3 & 0 \\ 0 & 3 \end{bmatrix} \begin{bmatrix} 6 \\ 4 \end{bmatrix} = \begin{bmatrix} 18 \\ 12 \end{bmatrix} = 3 \begin{bmatrix} 6 \\ 4 \end{bmatrix} = 3$ X

   2. so X = < 6 , 4 > is an eigenvector with an egienvalue of 3.

   3. we note something here ; so we try the following theorem.

III. Theorem. Any nonzero scalar multiple of an eigenvector X of a linear transformation T

         is also an eigenvector and each has the same eigenvalue.

   Proof.

   Given.  a  T  with eigenvector X ; so  T(X) = A X = $\lambda$ X with eigenvalue $\lambda$.

      1. let W = k X be another scalar multiple vector on X,

      2. then A W = A [ k X ] = k [ A X ] = k [$\lambda$ X ] = $\lambda$ [ k X ] = $\lambda$ W,

      3. so W = k X is also an eigenvector and both X & W have the same eigenvalue,$\lambda$.

IV. Problems. Given a T defined by the matrix A $= \begin{bmatrix} 0 & 1 \\ 1 & 0 \end{bmatrix}$

    A. Preform the A transformation on the

        vector X ($x_1$ ,$x_2$ ) where $x_1 \neq x_2$   to get vector W:

        1. W $= \begin{bmatrix} 0 & 1 \\ 1 & 0 \end{bmatrix} \begin{bmatrix} x_1 \\ x_2 \end{bmatrix} = \begin{bmatrix} x_2 \\ x_1 \end{bmatrix}$

        2. there is no scalar $\lambda$ to make

$$\lambda \begin{bmatrix} x_1 \\ x_2 \end{bmatrix} = \begin{bmatrix} x_2 \\ x_1 \end{bmatrix} : \quad x_1 \neq x_2$$

        3. so X is not a solution of the equation A X $= \lambda$ X .

        4. X($x_1$ ,$x_2$ ) ; $x_1 \neq x_2$   is not an eigenvector, and there is no eigenvalue.

        5. we have seen from the geometry that W ($x_1$ ,$x_2$ ) is the Reflection of the

vector X ($x_1$ ,$x_2$ ) through the  line $x_1 = x_2$  ; so the product  A  X($x_1$ ,$x_2$ ) defines

the reflection of X through the line $x_1 = x_2$  which we have seen not to be linear.

    B. Choose a vector X ($x_1$ ,$x_1$ ) that is on the line $x_2 = x_1$  . We see that

A X $= \begin{bmatrix} 0 & 1 \\ 1 & 0 \end{bmatrix} \begin{bmatrix} x_1 \\ x_1 \end{bmatrix} = \begin{bmatrix} x_1 \\ x_1 \end{bmatrix} = 1 \begin{bmatrix} x_1 \\ x_1 \end{bmatrix} = 1$  X       ;

        so X is an eigenvector with the eigenvalue of 1, and we note that the reflection of

        a vector on the line $x_2 = x_1$  is the vector itself.

11.3.  Note:

    1. In this and the last chapter we have seen how to define certain simple geometric

transformations involving vectors [stretching , reflections , rotations , ... ] by

setting up a matrix multiplication between a matrix, A, and the vector.

    2. The matrix A has been simple and easy to find, but this may not always be the

case.

    3. Our geometry is restricted to 2-space and  3-space.

    4. We need to do more work with the concept of  Eigenvectors and Eigenvalues.

11.4  Another approach to the evaluation of eigenvectors and eigenvalues.

  I. Given the special linear transformation : T(X) $= \lambda$ X as in the equation A X $= \lambda$ X,

        Take X $= \begin{bmatrix} x_1 \\ x_2 \end{bmatrix}$ as unknowns     &  A $= \begin{bmatrix} a_{11} & a_{12} \\ a_{21} & a_{22} \end{bmatrix}$ as knowns  ;

now even in 2-space it is not obvious how to solve A X $= \lambda$ X for X or for $\lambda$.

A.   Symbolic Form of our new approach.

   1. Given the equation   A  X  =  $\lambda$  X ,

   2. then  A  X  -  $\lambda$  X  = 0                           Or  [ A  -  $\lambda$ ] X  = 0

   3. we seem to have a subtraction between a nxn matrix A and a scalar $\lambda$; so consider this: from matrix theory we have seen that: $\lambda$ = $\lambda$ I. With that substitution the subtraction is meaningful because we have two matrices of equal orders.

   4. thus    [ A  -  $\lambda$  I ]  X  = 0.

   As an equation this is the best symbolic form of the  this special linear transformation. This is actually a system of n equations.

   B. Computational form for 2-space.

   1. So    A X = $\lambda$ X    =>  $\begin{bmatrix} a_{11} & a_{12} \\ a_{21} & a_{22} \end{bmatrix}$ $\begin{bmatrix} x_1 \\ x_2 \end{bmatrix}$ = $\lambda$ $\begin{bmatrix} x_1 \\ x_2 \end{bmatrix}$ = $\begin{bmatrix} \lambda\ x_1 \\ \lambda\ x_2 \end{bmatrix}$

   or $\left\{ \begin{array}{l} a_{11}\ x_1 + a_{12}\ x_2 - \lambda\ x_1 = 0 \\ a_{21}\ x_1 + a_{22}\ x_2 - \lambda\ x_2 = 0 \end{array} \right\}$

   2. thus the system of equations: $\left\{ \begin{array}{l} \{a_{11} - \lambda\}\ x_1 + \quad\quad a_{12}\ x_2 = 0 \\ \quad\quad a_{21}\ x_1 + \{a_{22} - \lambda\}\ x_2 = 0 \end{array} \right\}$

define both the eigenvectors X = < $x_1$ , $x_2$  >  and eigenvalues $\lambda$ , but with three forms of unknowns  { $x_1$  , $x_2$  , & $\lambda$ }  we need more work to find a solution..

II.

   A. More on the Symbolic form.

   1. From a previous chapter the homogeneous system of equations [A - $\lambda$ I ] X = 0

      also has non-trivial solutions iff   det | A - $\lambda$ I |  = 0

   2. This equation defines $\lambda$  for use in I.B.2. and is called the Characteristic Equation of A. Written as a function , f($\lambda$ ) = det | A - $\lambda$ I | , it is the Characteristic Polynomial of A of the nth degree. The matrix [ A - $\lambda$ I ] is the Characteristic Matrix of A. If a particular eigenvalue, say $\lambda$ , is a Root of Multiplicity  k  of the equation  det| A - $\lambda$ I | = 0  , then we say that $\lambda$  has Algebraic Multiplicity k  [ see example later ]. The Null Space of  [ A - $\lambda$ I ], labelled  $E_\lambda$ , is called the Eigenspace of A corresponding to the eigenvalue $\lambda$ ; so

   $E_\lambda$  =  Null[ A - $\lambda$ I ].

B. Computational form of this expanded material.

1. From A.1.    det $\begin{vmatrix} [a_{11} - \lambda] & a_{12} \\ a_{21} & [a_{22} - \lambda] \end{vmatrix}$ = 0

2. $[a_{11} - \lambda] [a_{22} - \lambda] - a_{12} a_{21} = 0$

   $a_{11} a_{22} - a_{11} \lambda - a_{22} \lambda + \lambda^2 - a_{12} a_{21} = 0$

3. $\lambda^2 - [a_{11} + a_{22}] \lambda + a_{11} a_{22} - a_{12} a_{21} = 0$

   note: there can two unequal real $\lambda$'s , two equal real $\lambda$'s, or two complex $\lambda$'s.

4. We now  solve this equation for $\lambda$ in terms of the given $a_{ij}$ ; then substitute each $\lambda$  into the system in step I.B.2.  to get each corresponding  $x_1$  &  $x_2$    .

4. While the algebra gets very complicated, we see how to expand this solution to spaces higher than 2.

III. Therefore, Given: a square matrix  A = $(a_{ij})$ nxn  and  X = $(x_{ij})$ nx1 that defines a linear transformation A X :

1. We find the Eigenvalues $\lambda$'s and the Eigenvectors X's of  A

   by solving  $[A - \lambda I] X = 0$ ;

2. it is also clear that we first find $\lambda$ from:

   det $| A - \lambda I | = 0$

   iff  each non-trivial solution $\lambda$  is an Eigenvalue of A,

3. and then each $\lambda$ gives a non-zero vector X from the system  $[A - \lambda I] X = 0$

IV. Nomenclature.

1. The set of all of the Eigenvectors of A that have the same Eigenvalue $\lambda_k$   plus the

   Zero Vector is called The Eigenspace $E_k$   of A with  respect to that $\lambda_k$   .

2. the Dimension of each $E_k$  is called the Geometric Multiplicity of that $\lambda_k$    . The

   Geometric Multiplicity is less than or equal to the Algebraic Multiplicity; the

   Geometric is the maximum number of Linearly Independent Eigenvectors that

   correspond to the Eigenvalue $\lambda_k$    .


11.5 Examples. [ note: while it is possible to use II. & III. to develop "shortcuts" and "devices" to find solutions,  working with the given and the basics seems to work just as well without extra memory work; as : ]

I. Problem. Given A = $\begin{bmatrix} 4 & 1 \\ 1 & 4 \end{bmatrix}$    Find Eigenvalues  & Eigenvectors  [note: student must verify the highlights listed below].

1. $\begin{bmatrix} 4 & 1 \\ 1 & 4 \end{bmatrix} \begin{bmatrix} x_1 \\ x_2 \end{bmatrix} = \lambda \begin{bmatrix} x_1 \\ x_2 \end{bmatrix}$ => $\begin{Bmatrix} [4 - \lambda] x_1 + & x_2 = 0 \\ x_1 + [4 - \lambda] x_2 = 0 \end{Bmatrix}$

2. $\begin{vmatrix} [4 - \lambda] & 1 \\ 1 & [4 - \lambda] \end{vmatrix} = 0$    => $\lambda^2 - 8\lambda + 15 = 0$

$\lambda = 3$ & $\lambda = 5$

3. $\lambda = 3$ into #1 reduces to : $\begin{Bmatrix} x_1 + x_2 = 0 \\ x_1 + x_2 = 0 \end{Bmatrix}$

   so multiple roots: let $x_1 = r$ [a scalar], then $x_2 = -r$,

   and the Eigenvectors $\begin{bmatrix} r \\ -r \end{bmatrix}$ have the Eigenvalue of 3 for all r's

4. $\lambda = 5$ into # 1 reduces to: $\begin{Bmatrix} -x_1 + x_2 = 0 \\ x_1 - x_2 = 0 \end{Bmatrix}$

   so multiple roots: let $x_1 = s$ , then $x_2 = s$,

   and the Eigenvectors $\begin{bmatrix} s \\ s \end{bmatrix}$ have the Eigenvalue of 5 for all s's

5. so $E_3 = \begin{bmatrix} r \\ -r \end{bmatrix}$    where r is any number

   $B_3 = \begin{bmatrix} 1 \\ -1 \end{bmatrix}$    is a Basis for $E_3$   [it Spans the space]

6. & $E_5 = \begin{bmatrix} s \\ s \end{bmatrix}$    where s is any number

   $B_5 = \begin{bmatrix} 1 \\ 1 \end{bmatrix}$    is a Basis for $E_5$

II. Problem. Given A = $\begin{bmatrix} 0 & 0 & 1 \\ 0 & 1 & 0 \\ 0 & 0 & 1 \end{bmatrix}$    Find the Eigenvalues and Eigenvectors.

1.
$$\begin{bmatrix} 0 & 0 & 1 \\ 0 & 1 & 0 \\ 0 & 0 & 1 \end{bmatrix} \begin{bmatrix} x_1 \\ x_2 \\ x_3 \end{bmatrix} = \begin{bmatrix} x_1 \\ x_2 \\ x_3 \end{bmatrix} \Rightarrow \begin{cases} -\lambda x_1 & + x_3 = 0 \\ [1 - \lambda] x_2 & = 0 \\ [1 - \lambda] x_3 = 0 \end{cases}$$

2.
$$\begin{vmatrix} -\lambda & 0 & 1 \\ 0 & [1-\lambda] & 0 \\ 0 & 0 & [1-\lambda] \end{vmatrix} = 0 \Rightarrow [-\lambda][1-\lambda][1-\lambda] = 0$$
$$\text{so: } \lambda = 0, \lambda = 1, \lambda = 1$$

3.  $\lambda = 0$ into # 1 : $x_3 = 0$, $x_2 = 0$, & $x_1$ drops out of the system so is arbitrary,

The Eigenvectors $\begin{bmatrix} r \\ 0 \\ 0 \end{bmatrix}$   r a scalar    all have the Eigenvalue $\lambda = 0$

4. $\lambda = 1$ into # 1 :   $-x_1 + x_3 = 0$  & all other equations become $0 = 0$;

so : $x_1 = x_3$   thus let $x_3 = t$  and then $x_1 = t$ , t a scalar;

now $x_2$   drops out therefore is also arbitrary; let  $x_2 = s$   where s is a scalar.

The Eigenvectors $\begin{bmatrix} t \\ s \\ r \end{bmatrix}$   t & s are scalars  all have the Eigenvalue  $\lambda = 1$

5. $E_0 = \begin{bmatrix} r \\ 0 \\ 0 \end{bmatrix}$   where r is any number

$B_0 = \begin{bmatrix} 1 \\ 0 \\ 0 \end{bmatrix}$   is a Basis for $E_0$

6. $E_1 = \begin{bmatrix} t \\ s \\ t \end{bmatrix}$   where t,s are any numbers

Basis:

a. if we let t = 1 & s = 0 we get one simple eigenvector $\begin{bmatrix} 1 \\ 0 \\ 1 \end{bmatrix}$

b. if we let t = 0 & s = 1 we get another simple eigenvector $\begin{bmatrix} 0 \\ 1 \\ 0 \end{bmatrix}$

c. now $c \begin{bmatrix} 1 \\ 0 \\ 1 \end{bmatrix} + k \begin{bmatrix} 0 \\ 1 \\ 0 \end{bmatrix} = \begin{bmatrix} 0 \\ 0 \\ 0 \end{bmatrix}$ iff $\left\{ \begin{array}{l} c \quad + k \cdot 0 \ = \ 0 \\ c \cdot 0 + k \quad = \ 0 \\ c \quad + k \cdot 0 \ = \ 0 \end{array} \right\}$ or c, k = 0

so the vectors < 1, 0, 1 > & < 0, 1, 0 > are Linear Independent; so this pair of vectors span the Eigenspace $E_1$ ; so a Basis for $E_1$ is the set

$$B = \left\{ \begin{bmatrix} 1 \\ 0 \\ 0 \end{bmatrix}, \begin{bmatrix} 0 \\ 1 \\ 0 \end{bmatrix} \right\}$$

11.6. Exercise. Find the Eigenvalues and Eigenvectors of the matrix A = $\begin{bmatrix} 1 & 2 \\ -1 & 4 \end{bmatrix}$

11.7 Solution of the Polynomial Equation P ($\lambda$) = 0.

1. If n = 2 , we will have no trouble in solving the equation, but if n= 3 we will have a little trouble in such solution. We will borrow a little bit from higher algebra to help with these solutions.

2. Given the equation : $a_n \lambda^n + a_{n-1} \lambda^{n-1} + a_{n-2} \lambda^{n-2} + ... + a_0 = 0$ where n ≥ 3 , and the $a_i$ are Integers, $a_n \neq 0$ , & $a_0 \neq 0$.

3. any rational root of this equation is a Divisor of $a_0$ ,

4. make a list of such divisors and try each one as a possible root.

5. if r is such that P (r ) = 0 , then divide P($\lambda$) by [$\lambda$ - r ]. The result Q($\lambda$) is a polynomial of one less degree where the roots of Q($\lambda$) = 0 are the same as other roots of P($\lambda$) = 0. Sometimes one can be very clever and factor P ($\lambda$) completely by grouping using the one factor [$\lambda$ - r].

6. If n = 3 , then Q($\lambda$) = 0 is of degree two and can easily be solved for the two remaining roots.

7. If n > 3 , then apply this method over and over picking up a root each time the degree of the Q($\lambda$) gets smaller. When the last Q($\lambda$) is a quadratic, finish the solution as in Elementary Algebra.

11.8. There are many concepts that use Eigenvectors and Eigenvalues; in this text we will only study a few.

11.9 Inverse Matrix.

Theorem. A square Matrix, A , is Invertible or Nonsingular  [ that is : has an Inverse ] iff no Eigenvalues are zero.

A. Choose n = 2  [ student must fill in missing stepts ]:

1. We have seen that if $A = \begin{bmatrix} a_{11} & a_{12} \\ a_{21} & a_{22} \end{bmatrix}$

2. $AX = \lambda X$ => $\begin{cases} [a_{11} - \lambda] x_1 + a_{12} x_2 = 0 \\ a_{21} x_1 + [a_{22} - \lambda] x_2 = 0 \end{cases}$

3. and $\det | A - \lambda I | = 0$ => $\lambda - [ a_{11} + a_{22} ] \cdot \lambda + [ a_{11} a_{22} - a_{12} a_{21} ] = 0$

4. If $a_{11} a_{22} - a_{12} a_{21} = 0$ , then $\lambda [\lambda - a_{11} - a_{22} ] = 0$ => $\lambda = 0$

5. If $a_{11} a_{22} - a_{12} a_{21} \neq 0$ , then in general $\lambda \neq 0$,

6. now $a_{11} a_{22} - a_{12} a_{21} = \det(A)$ ; so if $\lambda \neq 0$ , then $\det(A) \neq 0$

7. from previous $\det (A) \neq 0$ => A has an Inverse.

B. Trying to show this for $n \geq 3$ is very difficult ; so as in this first book we do not try for the general case under those circumstances.


11.10 Triangular Matrix.

Theorem. The Eigenvalues of either kind of Triangular Matrix
         are the Main Diagonal entries.

1. A triangular matrix is one where all of the entries below [or above] the main
   diagonal are zero, thus:

2. $\det (A) = a_{11} a_{22} a_{33} \ldots a_{nn} = \prod_{i=1}^{n} a_{ii}$

3. so $\det[ A - \lambda I ] = 0$ => $[a_{11} - \lambda] [a_{22} - \lambda] [a_{33} - \lambda] \ldots [a_{nn} - \lambda] = 0$

4. so the eigenvalues $\lambda$'s are: $a_{11}$ , $a_{22}$ , $a_{33}$ ,..., & $a_{nn}$ : the main diagonal entries.


Theorem. The Eigenvalues of a Diagonal Matrix are the Main Diagonal entries or elements.

[note: a diagonal matrix is first of all a triangular one ; so the proof is obvious ].

11.11 Eigenvalues and Linear Transformations.

I.

    1. Any linear transformation , T (V) = W can be written as T (V) = A (V) = W; so we can apply the concept of Eigenvalues of matrix A to T .

    2. Definition: If a linear transformation is special as in T (V) = $\lambda$ V ; V $\neq$ 0, then $\lambda$ is an Eigenvalue of T and V is the corresponding Eigenvector.

II. Example: Find the Eigenvalues/Eignevectors of the linear transformation

$$T <x, y> = < x + y , x + y >$$

1. Find A from T(X) = A X   [note: can show as either a vector or a matrix] as:

$$\begin{bmatrix} \{x + y\} \\ \{x + y\} \end{bmatrix} = \begin{bmatrix} a & b \\ c & d \end{bmatrix} \begin{bmatrix} x \\ y \end{bmatrix} \Rightarrow \begin{Bmatrix} x + y = a x + b y \\ x + y = c x + d y \end{Bmatrix}$$

so a = 1 , b = 1 , c = 1 , d = 1 ;   thus   $A = \begin{bmatrix} 1 & 1 \\ 1 & 1 \end{bmatrix}$

2. det ( A - $\lambda$ I ) = $\begin{vmatrix} 1 - \lambda & 1 \\ 1 & 1 - \lambda \end{vmatrix}$ = [ 1 - $\lambda$ ]$^2$ - 1 = 1 - 2$\lambda$ + $\lambda^2$ - 1

$$= \lambda^2 - 2\lambda = \lambda [\lambda - 2]$$

so det ( A - $\lambda$ I ) = 0   iff   $\lambda$ [$\lambda$ - 2] = 0

thus the eigenvalues are:   $\lambda$ = 0 , $\lambda$ = 2

3. now A X = $\lambda$ X ; thus $\begin{bmatrix} \{x + y\} \\ \{x + y\} \end{bmatrix} = \lambda \begin{bmatrix} x \\ y \end{bmatrix}$ or $\begin{bmatrix} \{x + y\} \\ \{x + y\} \end{bmatrix} = \begin{bmatrix} \lambda x \\ \lambda y \end{bmatrix}$

4. when $\lambda$ = 0 we have $\begin{Bmatrix} x + y = 0 \\ x + y = 0 \end{Bmatrix}$ ; let y = r then x = -r   r any scalar;

thus $\begin{bmatrix} - r \\ r \end{bmatrix}$ are the eigenvectors for the eigenvalue of $\lambda$ = 0 where r any scalar.

5. when $\lambda$ = 2 we have $\begin{bmatrix} x + y \\ x + y \end{bmatrix} = \begin{bmatrix} 2 x \\ 2 y \end{bmatrix}$ Or $\begin{Bmatrix} - x + y = 0 \\ x - y = 0 \end{Bmatrix}$ so y = x

let y = r then x = r

thus $\begin{bmatrix} r \\ r \end{bmatrix}$ are the eigenvectors for the eigenvalue of $\lambda$ = 2 where r any scalar.

11.12 Similar Matrices and Eigenvalues.

   I. Definition review. A matrix A is said to be Similar to a matrix B if there  exists a Nonsingular [ or Invertible ] Matrix P such that  $A = P B P^{-1}$. This Similarity is symbolized by  $A \sim B$.

   II. Theorem. If the square matrices A and B are Similar,

           Then A and B have identical sets of Eigenvalues.

      Proof.

      1. given $A \sim B$  ,so by the definition there is a Nonsingular square matrix P such that  $A = P A P^{-1}$   ---   note that $P P^{-1} = I$,

      2. the eigenvalue $\lambda$ of A is determined by  $\det( A - \lambda I ) = 0$

thus:   $\det( A - \lambda I ) = \det ( P B P^{-1} - \lambda I ) = \det ( P B P^{-1} - \lambda P P^{-1} )$

              $= \det ( P B P^{-1} - P \lambda P^{-1} ) = \det ( P [ B P^{-1} - \lambda P^{-1} ] )$

              $= \det ( P [ B - \lambda ] P^{-1} ) = \det ( P [ B - \lambda I ] P^{-1} )$

              $= ( \det P ) ( \det [ B - \lambda I ] ) ( \det P^{-1} )$

      3. P has an inverse so $P P^{-1} = I$

       so      $\det ( P P^{-1} ) = \det I$

      $\det ( P )  \det ( P^{-1} ) = 1$

             or  $\det ( P ) = 1/\det( P^{-1} )$

              $[\det ( P )] = ( \det P )^{-1}$

      4. into #2 :   $\det ( A - \lambda I ) = [ \det P ] [ \det ( B - \lambda I ) ][ ( \det P^{-1} ) ]$

                        $= [ (\det P ) ( \det P^{-1} )] [ \det ( B - \lambda I ) ]$

                        $= I [ \det ( B - \lambda I ) ]$

                        $= \det ( B - \lambda I )$

      5.  to find $\lambda$  let either $\det( A - \lambda I ) = 0$  or $\det ( B - \lambda I ) = 0$

      6. so, A & B have the same Characteristic Polynomial; thus they have the same

         Eigenvalues.

      7. note: Similar Matrices form an Equivalent Class  [ relative to

         Eigenvalues/Eigenvectors ] ; they are Equivalent to each other.

I. A review of the DERIVATIVE concept where the functions, F(x), f(x) , & g(x)

   are all Continuous for x ∈ R and all Limits [lim] are for h→0.

 1. Definition of a Derivative :  The Derivative of the function y = F(x) is:

   $D\{F[x]\} \equiv dy/dx \equiv F'(x) = \lim_{h \to 0} \{F(x + h) - F(x)\}/h$   if the limit exists.

2. The Addition formula for Derivatives:

   $D[f(x) + g(x)] = \lim_{h \to 0} \{ [f(x + h) + g(x + h)] - [f(x) + g(x)] \}/h$

   $= \lim_{h \to 0} \{ f(x + h) + g(x + h) - f(x) - g(x) \}/h$

   $= \lim_{h \to 0} \{ [f(x + h) - f(x)] + [g(x + h) - g(x)] \}/h$

   $= \lim_{h \to 0} \{ [f(x + h) - f(x)] \}/h + \lim_{h \to 0} \{[g(x + h) - g(x)] \}/h$

   $D[f(x) + g(x)] = D[f(x)] + D[g(x)]$

   In words: The Derivative of a Sum is the Sum of the Derivatives.

3. The Scalar Product formula for  Derivatives.

   $D[ c\, f(x) ] = \lim_{h \to 0} \{ c\, f(x + h) - c\, f(x) \}/h$      , c   a constant

   $= \lim_{h \to 0} \{ c\, [ f(x + h) - f(x) ] \}/h$

   $= c\, \{ \lim_{h \to 0} \{ f(x + h - f(x) \}/h$

   $D[ c\, f(x) ] = c\, D[ f(x) ]$         , c a constant

    In words: The Derivative of a Constant times a function of x is the constant times the derivative
the function.

II. IRRATIONAL NUMBERS.

When we try to finish The Addition Identity Theorem [1.3. V.] by extending the domain to include irrational numbers and thus cover the entire Real Number system, we run into some serious problems. The study of irrational numbers is not a simple thing, but without digging too deeply into Number Theory we can make some inroads to make what we need seem reasonable. For instance:

1. We know that $\sqrt{2}$ is an irrational number [another word for not-rational].

Note the following proof:

a. assume $\sqrt{2}$ is rational; or: $\sqrt{2} = a/b$ , $a,b \neq 0$, $a,b \in I$, and a & b have no common factors — that is a/b has been reduced to lowest terms,

b. by definition of square root this means $2 = a^2/b^2$ ,

c. or $2b^2 = a^2$ ; thus $a^2$ is even which means "$a$" is even,

{ 2x represents all even numbers and $[2x]^2 = 4x^2$ ; so an even number squared is even.

2x + 1 represents all odd numbers and $[2x + 1]^2 = 2[2x^2 + 2x] + 1$; so an odd number squared

is odd. Integers are either even or odd. }

d. let a = 2d thus in step c. $2b^2 = 4d^2$ or $b^2 = 2d^2$ ;

so $b^2$ is even ; so $b$ is even,

e. thus a & b both have a factor of 2; so we have a contradiction to step a,

f. $\sqrt{2}$ is Not Rational and therefore Irrational.

2. let $\sqrt{2} = r$ an irrational number; by squaring we can show that the following nested intervals are true:

| | | | | |
|---|---|---|---|---|
| 1 | $< \sqrt{2} < 2$ | on squaring we get : 1 | $< 2 < 4$ | which is true, |
| 1.4 | $< \sqrt{2} < 1.5$ | on squaring we get : 1.96 | $< 2 < 2.25$ | which is true, |
| 1.41 | $< \sqrt{2} < 1.42$ | on squaring we get : 1.9881 | $< 2 < 2.0164$ | which is true, |
| 1.414 | $< \sqrt{2} < 1.415$ | on squaring we get : 1.999396 | $< 2 < 2.002225$ | which is true, |
| 1.4142 | $< \sqrt{2} < 1.4143$ | on squaring we get : 1.9999616 | $< 2 < 2.0002445$ | which is true, |

and so on

3. It is clear that as we move down this array, we are getting better and better approximations to a rational "value" of this irrational number [note: 1.41 = 141/100; so 1.41 is a rational number]. It seems reasonable then that an irrational number has been defined in other mathematics as a Decimal Number with an infinite number of decimals places.

4. The above describes a Nested Sequence of Intervals as in the following diagram:

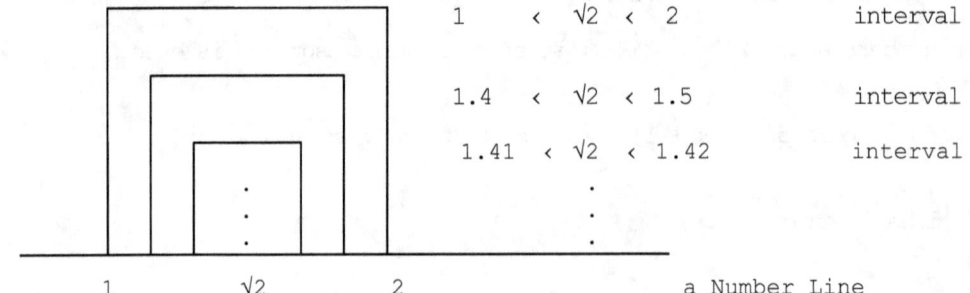

| | | |
|---|---|---|
| 1 | $< \sqrt{2} < 2$ | interval |
| 1.4 | $< \sqrt{2} < 1.5$ | interval |
| 1.41 | $< \sqrt{2} < 1.42$ | interval |

a Number Line

In finding a definition of Irrational numbers we can choose the numbers on the left side of the intervals and let $r_1 = 1.4$ , $r_2 = 1.41$ , $r_3 = 1.414$ , $r_4 = 1.4142$ , and so on; and set up the nested sequence $\{ r_1 \} = r_1 , r_2 , r_3 , \ldots , r_n$ .

Then it seem reasonable that $\lim_{n \to \infty} r_n = \sqrt{2}$

5. It also seems reasonable that we can find a nested sequence for each different kind of irrational number with which we have worked. There are other methods of defining irrational numbers, and they agree in essence with this definition of an irrational number.

6. Therefore, in all cases we do accept that for every irrational number r, there is a nested sequence as in $\{ r_n \} = r_1 , r_2 , r_3 , \ldots, r_n$ where

$$r = \lim_{n \to \infty} r_n$$

# INDEX

abstract  12
Abstract Vector Space  34
addition identity  2
addition property  7
arrow  17

basis  91, 187

Cauchy-Schwarz Inequality  60
centroid  48
characteristic equation  198
complex numbers  91
Cramer's rule  162
cycloid  50

determinant 155, 159
   cofactor  158
   minor  158
   properties  160
dihedral angle  76
dimension  41
directional cosines  66

eigenvalues/eigenvectors  195
epicyloid  51
equal class  14
equation of a line  79
equation of a plane  75
equivalent class  14

function  9

Gauss elimination  142

homogeneity property  7

kernal  43
Kronecker Delta  43

linear combination  39
linear function  6
linear independent  40
linear property  6, 10
linear transformation  43, 179

matrix  94
  addition  98
  adjoint  167
  augmented matrix  117
  basis  101
  diagonal  104
  echelon form  123
  equal  96
  equivalent  118
  graphic product  110
  inverse  114,133,203
  negative  97

  product  102
  properties  111-114
  rank  126
  row/column  95
  row operations  117
  scalar  104
  scalar product  99
  sigma symbol  104
  similar  205
  singular  136
  submatrix  96
  subtraction  99
  transpose  102
  triangular  107, 203
  unit  97
  zero  97
multiplicity 198

n-tuple  38

one-to-one correspondence  15
operation relation  10

projection  19

skew lines  87
space  16
span  41
subspace  42
system if equations  116
  echelon method  142
  homogeneous  149
  solutions  141

three-space  64
transformation relation  10
triangle inequality 61, 63
triple scalar product  85
triple vector product  81

vector  17
  algebraic addition  22, 65
  basic theorems  23, 26
  collinear  31
  cross product  67, 69, 71
  directional cosines  66
  distributive for cross product  71
  dot product  54, 65
  equal  17
  free  17
  geometric addition  21
  line segment  18

INDEX   (continued)

vector
   magnitude  19, 65
   negative  24
   normalizing 44
   ordered pair  18
   parallel  17
   position  18, 64
   scalar product  24, 65
   unit  29
   zero  24